A DICTIONARY OF

English Plant Names

(and Some Products of Plants)

GEOFFREY GRIGSON

ALLEN LANE

For *La grande cuisinière*
(sur une échelle)

Contents

Foreword

The names in this glossary are the common English ones for plants (and products of plants) from trees to seaweeds – plants of the wild, plants of the farm and the vegetable garden and the kitchen, of the greengrocer's and the chemist's shop, plants of the flower border and the greenhouse, not excluding plants of literature such as the *Long Purples* carried by Ophelia or the *Hebenon* whose juice was poured into the ear of Hamlet's father, or mythical plants such as the *Yggdrasil* under which the gods of the Norsemen sat in council or the *Upas* of Java which spread over the skulls and ribs of the criminals who had succumbed to its poison. As for products, entries will be found for *Opium* as well as *Poppy*, *Cider* as well as *Apple* (and such particular apples as *Cox's Orange Pippin* or the *Quarrenden*), *Wine* as well as *Vine*, *Absinthe* as well as *Wormwood*, *Marmalade* as well as *Quince*. And it seemed to me that there is a good case for including, let's say, the *Gopherwood* of the Ark no less than the *Gingko*.

An English name is taken to be one in familiar usage – such usage having so often, for example, captured the 'Latin' of botanical nomenclature, when it happens to offer an attractive symbol, *Zinnia*, for example, or *Dahlia*, from the botanists Zinn and Dahl.

In strict etymology many of the names included are of anything but recondite derivation. Yet behind their obviousness there may be facts worth explaining, which lead back across Europe, or further still, into antiquity. If it is intriguing to know that we are using a word familiar to the Aztecs when we ask for an *Avocado*, it has its fascination to discover that the Elizabethans were the first to use the composite name *Seville Orange*.

The entries give their information in the following order: first the English plant name. Then in brackets the scientific name of the species particularly or usually indicated by the English name; other English synonyms; and the date when the species was introduced into England, in Arabic numerals. Beyond the bracket comes in Roman numerals the century in which the plant name is first on record (no such date is given when a name is stated to go back to Old English, i.e. English from the eighth century till about 1100). The century date is followed sometimes by the name of the man by whom the plant-name was invented or first recorded. Last of all the name, where possible, is historically or etymologically explained.

Since it is confusing and irritating to have to refer to a code whenever one looks up a name, abbreviations have been kept to a minimum of the obvious. So have special signs or technical terms.

However, in expansion of brevities in the pages which follow, it will help to add a list of language dates and explanations, less known languages from which plants and plant products are named, various literary sources, and some of the English writers who first recorded or devised many of our plant names.

ACCADIAN. The ancient Semitic language of the Assyrians and Babylonians.

ANGLO-NORMAN. The dialect of French spoken by the Normans in England.

ARAWAK. Language of northern South America and formerly much of the West Indies, contributing plant names via the expeditions of Columbus, and so through Spanish.

CANTONESE. Chinese of S.E. China, Canton, Hong Kong.

CARIB. Language of northern South America which Columbus found in the eastern islands of the Caribbean.

COGNATE. Used of words or names related to other names or words by descent from a common ancestor.

CULPEPER. The herbalist Nicholas Culpeper (1616–1654). Author of the racily worded herbal *A Physicall Directory*, 1649; of which later editions (1653, etc.) were re-named *The English Physitian Enlarged*.

DIOSCORIDES. The army physician Dioscorides Pedanius, a

Greek who compiled about AD 60 (in Greek) his *Hules Iatrikes* (or *De Materia Medica*), a herbal or account of the medicinal plants of Asia Minor, which was of unquestioned authority in the revival of plant studies in the sixteenth century.

DODOENS (Dodonaeus). Rembert Dodoens (1517–1585), great Flemish botanist. Many English plant names come from the Dutch of his *Cruydeboeck*, 1554, through Henry Lyte's translation (from the French edition), *A Niewe Herball, or Historie of Plantes*, 1578, and from his *Pemptades*, 1583, a translation of which is the substance of Gerard's famous *Herball*, 1597.

DRAVIDIAN. Family of pre-Indoeuropean languages of India, including Malayalam and Tamil.

ELYOT. Sir Thomas Elyot (c. 1490–1546), English scholar and humanist, introducer of the word 'democracy'. Elyot 1538, etc., refers to editions of *The Dictionary of Syr Thomas Elyot*.

EVELYN. John Evelyn (1620–1706), diarist and gardener, who Englished a number of French names for vegetables, etc. Translator of *The French Gardiner*, 1658. Author of *Sylva, or A Discourse of Forest-Trees*, 1664; *Kalendarium Hortense*, 1664; *Acetaria. A discourse of sallets*, 1699.

FRISIAN. One of the West Germanic group of Indoeuropean languages, close to English, with which it shares many plant names. The language of the original homelands of the English.

GERARD. John Gerard (1545–1607), London apothecary. Author, or rather editor, of *The Herball or Generall Historie of Plantes*, based on Dodoens (see above). Many plant names first appear in the *Herball*, which was evidently read by Shakespeare. The editions are referred to as Gerard 1597 and Gerard 1633.

GERMANIC. The Germanic languages, one of the sub-families of the Indoeuropean languages, which includes North Germanic and West Germanic, explained below.

HINDI. One of the Indoeuropean languages of modern India, from which some plant names have passed direct into English.

INDOEUROPEAN. The great family of languages which includes Sanskrit, Persian, Greek, Latin, the Celtic languages, the Romance languages, the Germanic and the Slavonic languages.

LATE LATIN. Latin from AD 200 to 600.

LOW GERMAN. The German of North Germany, from Old Saxon to the modern Plattdeutsch dialects.

LOW LATIN. Spoken Latin – in contrast to the written classical Latin – of the later Roman empire to the end of the fifth century AD, which develops and diverges into the Romance languages.

LYTE. Henry Lyte (c. 1529–1607), translator and adapter for English use of the *Cruydeboeck* of Dodoens (see above). He lived at Lyte's Cary House, Charlton Mackrell, in Somerset, and was adept at devising plant names equivalent to names in Dutch or German. His *A Niewe Herball, or Historie of Plantes* is referred to as Lyte 1578.

MALAY. One of the Austronesian family of languages contributing plant names from Malaysia and the East Indies, especially through Dutch and Portuguese.

MALAYALAM. Dravidian, pre-Indoeuropean language spoken on the west coast of India, plant names from which have come to us often through Portuguese.

MANDARIN. Northern and literary Chinese.

MEDIEVAL LATIN. Latin from AD 600 to 1500.

MILLER. Philip Miller (1691–1771), gardener in charge of the Chelsea Physic Garden, author of the great gardening handbook of the eighteenth century *The Gardeners Dictionary*, cited as Miller 1731, etc.

MOZARABIC. Pertaining to the Mozarabic population of Spain, i.e. the Spanish Christians under the Moorish kings in Spain.

NAHUATL. The language of the Aztec empire, from which plant names have come down through the Spanish.

NORTH GERMANIC. The Germanic group comprising the Scandinavian languages.

OLD ENGLISH. English until about 1100.

OLD FRENCH. French from the ninth century to the end of the thirteenth century.

OLD HIGH GERMAN. German before the twelfth century.

OLD NORSE. The North Germanic languages of Norway, Denmark, Sweden, Iceland, England, etc., until about 1350.

OLD SAXON. The Low German spoken by the Saxons of N.W. Germany until the twelfth century.

Oxford English Dictionary, 1933. The prime source for the history, etymology and first occurrence of English words; and for subsequent dictionaries and glossaries (including this one).

PARKINSON. John Parkinson (1567–1650), apothecary and botanist, author of *Paradisi in Sole Paradisus Terrestris*, the first handbook of English pleasure gardening, and of *Theatrum botanicum . . . an Herball of a Large Extent*, cited respectively as Parkinson 1629 and Parkinson 1640. Parkinson recorded or devised names for many new plants from North America.

PERSIAN. Language of the Iranian sub-group of the Indoeuropean, source and transmitter to the West of plant names, from India, and through Arabic.

PLATTDEUTSCH. The Low German dialects of Northern Germany.

PLINY. Gaius Plinius Secundus (AD 23–79). Roman contemporary of Dioscorides (see above). Several of the thirty-six books of his encyclopedic *Naturalis Historia* are devoted to plants. These were known to every herbalist of the sixteenth century.

PRAKRIT. Ancient Indian regional languages forming with Sanskrit the Indo-aryan sub-group of the Indoeuropean languages.

PRATT. Anne Pratt (1806–1893), Victorian popularizer of English field botany, who both invented and recorded for the first time many plant names. Scarlet Pimpernel was one of her happy coinages. Editions of her *Flowering Plants and Ferns of Great Britain* are cited as Pratt 1855, etc.

PROVENÇAL. One of the Romance languages descended from Low Latin. An intermediary, from the geographical position of Provence, in the transmission of Latin and Italian plant names.

QUECHUA. The language of the Inca empire, contributing plant names through Spanish.

RAY. John Ray (1627–1705), great English naturalist and botanist, with a liking for words and names. *Catalogus Plantarum Angliae*, 1670; *Historia Plantarum*, vol. 1, 1686, vol. 2, 1688, vol. 3, 1704; *Synopsis Stirpium Britannicarum*, 1690.

ROMANCE, ROMANIC. The languages developed from Low Latin – Italian, Provençal, French, Spanish, Portuguese, etc.

SANSKRIT. The classical Indoeuropean language of ancient India.

STOCKHOLM MEDICAL MS. English manuscript herbal of the first half of the fifteenth century in the Royal Library, Stockholm, the basis of *Agnus Castus: A Middle English Herbal*, edited by Gösta Brodin, 1950.

SUMERIAN. Ancient pre-semitic language of Mesopotamia, which contributed some plant names through Accadian.

TAGALOG. One of the Austronesian languages, and the national language of the Philippines.

TAINO. See Arawak.

TAMIL. One of the Dravidian family of pre-Indoeuropean languages of modern India.

TUPI-GUARANI. Indian language of Brazil which has contributed plant names through Portuguese.

TURNER. William Turner (c. 1510/15–1579), humanist, reformer, and the father of English plant studies. Translated many plant names into English, and recorded or devised others. He is often cited.

Turner 1538: *Libellus de re herbaria nova.*

Turner 1548: *The names of herbes in Greke, Latin, Englische, Duche and Frenche wyth the commone names that Herbaries and Apotecaries use*, 1548.

Turner 1551, 1562 or 1568: *A new Herball* (published in three parts).

VULGAR LATIN. See Low Latin.

WEST GERMANIC. The group of the Germanic sub-family of

Indoeuropean languages which is made up of German, Low German, Frisian, Dutch, Flemish and English. Plant names common to these languages include many for the crops and weeds of agriculture, and for trees.

Note: An asterisk (*) before a plant name or form in the etymologies signifies that it is conjectural (as with some Old English plant names evidenced only in the names of places).

Acknowledgements

For help on particular points the compiler is grateful to Miss Tao Tao Liu of the Oriental Institute, Oxford; the Department of Printed Books in the National Library of Wales; the Librarian of the Royal Botanic Gardens, Kew; Professor Edward Schafer, of the University of California, Berkeley; Professor Kenneth Jackson, of Edinburgh University; and the Deputy Secretary of the Royal Horticultural Society.

A

AARON'S ROD (*Verbascum thapsus*, Mullein; also a name for *Agrimonia eupatoria*, Agrimony, and *Solidago virgaurea*, Golden Rod). XIX century. The tall erect raceme of *V. thapsus* suggested the rod of Levi on which Aaron's name was written and which 'budded and brought forth buds, and bloomed blossoms' (*Numbers* xvii, 3–8).

ABELE (*Populus albus*). XVII century. 'There is a something finer sort of White Poplar, which the Dutch call abele, and we have of late much of it transported out of Holland' (Evelyn 1664). Dutch *abeel*, through French, from medieval Latin *albellus*, the 'little white (one)', on account of the whitish underside of the leaves. The name *Dutch Arbel* has been recorded.

ABSINTHE (alcoholic drink). XIX century. From the French *absinthe*, drink flavoured with the plant *absinthe*, Wormwood (*Artemesia absinthium*), from Latin *absintium*, from Greek *apsinthion*, Wormwood, a loan-word from Persian. *The Oxford English Dictionary*'s first quotation for absinthe (1854) comes from *The Newcomes* by Thackeray.

ACACIA (*Robinia pseudoacacia*, of North American origin, the False Acacia, or Locust-tree, introduced in 1640). XVII century. 'The Acacia, together with that from Virginia, deserves a place among our avenue trees' (Evelyn 1664), referring both to the true acacias, which are wattles or mimosas, and the tree from the eastern U.S.A. now planted and naturalized in many parts of the world. Dioscorides used the Greek *akakia* (Pliny's *acacia*) for the Egyptian Pod-Thorn or Gum Arabic Tree, *Acacia nilotica*, a name presumably related to Greek *ake*, 'a point or thorn' (cf. Latin *acer, acus, acutus*), the tree with sharp thorns.

I

ACANTHUS (*Acanthus mollis*, Brank-ursine, from Southern Europe, introduced in the 16th century, the leaf-shape of which is formalized in the Corinthian capital). XVII century. From Latin *acanthus*, from Greek *akanthos*, *ake*, 'a prickle' or 'thorn', plus *anthos*, 'flower', in reference to the spiny bracts of the flower. See also *Bear's-breech*, *Brank-ursine*.

ACONITE (*Aconitum anglicum*, the Monk's-hood of gardens; used also for *Eranthis hyemalis*, the Winter Aconite). XVI century, Lyte, 1578. From the Greek *akoniton*, of unknown significance, applied by classical authors chiefly to the yellow-flowered *A. anthora*.

ADAM'S NEEDLE (*Yucca filamentosa* from the U.S.A., introduced 1675). Garden name of XIX century, the narrow leaves with their curly marginal threads having recalled *Genesis* iii, 7, after Adam and Eve had tasted the forbidden fruit, 'And the eyes of both of them were opened, and they sewed fig leaves together, and made themselves aprons.' See *Spanish Bayonet*, *Yucca*.

ADDER'S TONGUE (*Ophioglossum vulgatum*). XVI century, Lyte 1578, Englishing the *ophioglossum*, *Natterzünglein*, *Natterzung*, *Schlangenzünglin*, etc., of 16th century German authors from Bock and Paracelsus. Descriptive of the fertile spike of this small fern emerging from the blade which sheathes it.

AGAR-AGAR (gelatinous extract of various red seaweeds, especially of the genera *Gelidium* and *Gracilaria*). XIX century. From the Malayan *agar-agar*. (Agar-agar was first imported from Ceylon, as *Ceylon Moss*.)

AGAVE (various species of the genus *Agave*, including especially *A. americana*, the Century Plant or American Aloe, native of tropical America, introduced 1640). XIX century. The noble inflorescence suggested Greek *Agave*, a name of one of the Amazons and of other mythological characters, by origin the feminine of *agauos*, 'noble', 'illustrious'.

AGNUS CASTUS (*Vitex agnus-castus*, Chaste Tree, from Southern Europe, introduced in the 16th century). XIV century. By the

2

1. Acanthus, Bear's-breech, Brank-ursine (Matthiolus, *Commentarii*, 1565)

Agrimony

Greeks called *lugos* or *agnos*. At the Thesmophoria, the festival of Demeter Thesmophoros (Demeter the Law-Giver), Greek women sat themselves on the ground on branches of *agnos* to induce fertility, and refrained during the festival from coupling with their husbands. By association with Greek *hagnos*, 'chaste', the *agnos* became a symbol of chastity: Athenian wives, according to Pliny, strewed *agnos* leaves on their beds at the Demeter festival. Pliny's names for this shrub were *vitex* and – reduplicatively – *agnus castus* (*castus*, 'chaste'). Christian symbolism added to its virtue the chaste innocence of the *Agnus Dei*, the Lamb of God (though in England, until the true *Vitex agnus-castus* was introduced to gardens late in the 16th century, the name Agnus Castus, along with some of its virtues, was given to *Hypericum androsaemum*, Tutsan, q.v.). See *Chaste Tree*.

AGRIMONY (*Agrimonia eupatoria*). XIV century, *egremounde*; XV century, *agrimony*. Corrupted, and transferred from a different species, the *argemone* of the Greeks, possibly the small inconspicuous poppy, *Papaver argemone*, now a crop weed through most of Europe. *Argemone* derived from Greek *argemon*, the condition known as albugo – white speck on the cornea of the eye (*arges*, white), for which the Greek plant was prescribed. *Argemone* became Latin *argemonia*; this was misread in Pliny's *Natural History* as *agrimonia*, and applied in the early Middle Ages to the unpoppylike *A. eupatoria*.

AILANTHUS (*Ailanthus altissima*, from China, introduced 1751). XIX century. English use of the botanists' Latin, earlier *ailantus*, from *ailantho*, 'tree-heaven', the name used in Amboina, Indonesia, from *A. moluccana*. Cf. T. S. Eliot in 'The Dry Salvages': 'the rank ailanthus of the April dooryard'.

ALDER (*Alnus glutinosa*). Old English *alor*, *aler*, with cognates in other Germanic languages. From the same root as Old High German *elo*, *elawer*, 'reddish-yellow', the alder perhaps deriving its name from the startling reddish-yellow of its fresh-cut timber.

4

2. Agrimony (Fuchs, *De Historia Stirpium*, 1542)

ALE (alcoholic drink made from barley, with bitter plant flavourings). Old English *alu, ealo,* from an Indoeuropean base meaning 'bitter'.

ALECOST (*Chrysanthemum balsamita,* Costmary, of Western Asian origin, introduced by the 16th century). XVI century. *ale + cost,* from Latin *costum,* Greek *kostos* (an Indian plant whose scented root was used in compounding perfumes), from Hebrew *qōsht,* from Sanskrit *kusthah.* Alecost, with its mint-scented leaves, was grown especially for the flavouring and maturing of ale, in the days of extensive home-brewing. See *Costmary.*

ALEXANDERS (*Smyrnium olusatrum,* native of Southern Europe, grown from ancient times as a vegetable. Superseded in England in 18th century by celery). Old English *alexandre;* 13th–14th century *alexandrum, alysaundre;* from medieval Latin (*petroselinum*) *Alexandrinum,* 'rock-parsley' or 'rock-celery of Alexandria'. The meaning of *petroselinon,* the usual Greek name for Alexanders, and the instructions for growing it which Columella gives in his farming manual *De Re Rustica* (1st century AD), accord with its persistent habit in Western Europe, where it favours the neighbourhood of rock and stone walls, is not particular about soil, and as Columella remarks, goes on for ever once it is established.

ALFALFA (*Medicago sativa,* Lucerne, from the Mediterranean region and Western Asia, introduced in the late Middle Ages). XIX century. Spanish *alfalfa,* from Arabic *al* (the definite article) + *fasfasah.*

ALGAE (order of moisture-loving cryptograms, including the seaweeds). XVI century, Turner 1568, *alga,* for seaweed. Plural of Latin *alga,* 'sea-weed', which had the sense of something winding and binding, as in *ligare, alligare,* 'to bind' (cf. *ligament*).

ALKANET (now *Pentaglottis sempervirens,* Evergreen Alkanet). XIV century as a dye; XVI century for the dye plant. Medieval dyers gave the plant name Alkanet to the imported red dye-stuff obtained from *Alkanna tinctoria* of S.E. Europe, the true *Alkanet,* from Spanish *alcaneta,* diminutive of *alcana,* from Arabic *al-henna,* 'the *henna*', i.e. the shrub *Lawsonia inermis*, source of the henna dye: *alcaneta* was thus the 'little henna', in contrast to the henna shrub. Now wild in S.W. England, *P. sempervirens* also provides a red dye, and was possibly introduced and cultivated as a substitute for *Alkanna tinctoria*.

ALKEKENGI (*Physalis alkekengi*, Winter Cherry, Chinese Lantern, from S.E. Europe, introduced by 1548). XIV century. From the medieval Latin *alkakenge*, from the Arabic *al-kākandj*, *al*, 'the', + Persian *kākunadj*, 'nightshade'.

ALLELUIA (*Oxalis acetosella*). See *Wood-sorrel*.

ALLGOOD (*Chenopodium bonus-henricus*). XVI century, Lyte 1578. Equivalent of the medieval Latin *tota bona*, one of the names by which this plant was known to Tudor botanists. It was eaten as a vegetable like spinach; leaves, stems, inflorescence and all. See *Good King Henry, Mercury*.

ALLSEED (*Chenopodium polyspermum*). XVIII century. Translation of the 17th century botanists' *polyspermum*, *polysporon*, adapted from the Greek *poluspermos*, *polusporos*, 'many-seeded', a name given to the uncommon *C. polyspermum* which rather obviously displays its brown, glistening seeds.

ALLSPICE (spice from the dried berries of *Pimenta officinalis*, native of the West Indies and Central America). XVII century. The berries combine the flavour of cloves, nutmeg and cinnamon.

ALMOND (stone-fruit of the tree *Prunus amygdalis*, native of Western Asia, introduced in the 16th century). XIII century, for the fruit; XVII century for the tree. From the Old French *amande*, from Low Latin *amandula*, from Latin *amygdala*, Greek *amugdale*, from Hebrew *megedh el*, 'sacred fruit'. French *amande* became *almond* in English from the influence of *al*, the Arabic definite article. The Low Latin *amandula* gave Italian *mandorla*, for almond, familiar also as the almond shape (a symbolic intersection of the circles of heaven and earth, spirit and matter) enclosing God, Christ, or the Virgin Mary, in medieval art.

7

ALOE, ALOES (*1.* aromatic diseased wood of species of *Aquilaria*, especially *A. agallocha*, from Indochina, used in incense, etc.; *2.* drug from species of the genus *Aloe*, especially the Indian *Aloe vera*). In the first sense x century, Old English *aluwe*; in the second sense XIV century. From the Latin *aloe*, Greek *aloe*, from Hebrew *ahaloth*, from Sanskrit *agaru* ('the wood').

ALPENROSE (*Rhododendron ferrugineum*, from the mountains of Central Europe, introduced 1752). XIX century. German 'alpine rose'.

ALSIKE, ALSIKE CLOVER (*Trifolium hybridum*, Swedish Clover, from Atlantic and Mediterranean coastal areas of Europe, introduced as a forage crop 1777). XIX century (1852). From Alsike, in Sweden, between Stockholm and Uppsala, recorded by Linnaeus as a place where *T. hybridum* grew.

AMARANTH (mythical flower, as in Milton's *Paradise Lost*; and species of the genus *Amaranthus*, which includes the familiar Love-lies-bleeding). From Greek *amaranton*, 'not fading', 'everlasting'; Latin *amarantus*. The *amaranton* of the Greeks, which seems to have been one of the golden-flowered species of the genus *Helichrysum*, they also called *helichrusos* and *chrusanthemon*. As an everlasting it symbolized immortality, was represented on tombs, and was woven into chaplets. The *locus classicus* for the mythical amaranth as a deathless flower is Milton's *Paradise Lost*, iii, 351ff., where the angels throw down their crowns 'inwove with Amarant and Gold' –

> Immortal Amarant, a Flour which once
> In Paradise, fast by the Tree of Life
> Began to bloom, but soon for Mans offence
> To Heav'n remov'd where first it grew, there grows
> And flours aloft shading the Fount of Life . . .
> With these that never fade the Spirits Elect
> Bind thir resplendent locks inwreath'd with beams.

One's immediate mental coupling of amaranth with moly (q.v.) is due to *The Lotos-Eaters* of Tennyson, 'Propt on beds of amaranth and moly'. See also *Chrysanthemum, Everlasting*.

ANANAS (Pine-apple, fruiting inflorescence of *Ananas comosus*, from tropical America, introduced in the 17th century). XVII century. From the Spanish *ananas*, *anana*, from Portuguese *ananás*, from the Tupi-Guarani *nana*, *anana*, *ananas*.

ANCHUSA (garden species of the genus *Anchusa*, especially forms of the Mediterranean *A. azurea*). XX century. English use of the Latin generic name, from the Greek *agkhousa*, the name for the true Alkanet, *Alkanna tinctoria* (see *Alkanet*, above).

ANEMONE (species of the genus *Anemone*, including the Wood Anemone, Japanese Anemone, Poppy Anemone, etc.). XVI century. From Latin *anemone*, Greek *anemone*, by which the Greeks chiefly understood *A. coronaria*, the Poppy Anemone, of which there are now so many garden forms. According to Greek myth, the first anemones, brilliantly sanguine, grew from the blood which fell from Adonis when he was killed by the boar on Mount Lebanon, or else from the tears Aphrodite dropped as Adonis, her lover, lay dead on the ground. The Ancients at the same time interpreted *anemone* as a derivation from *anemos*, 'wind'. Pliny says, 'The flower never opens unless a wind is blowing; and from this it takes its name.' The myth and the explanation are combined by Ovid in the *Metamorphoses*, X. The goddess sprinkles nectar on the blood and within an hour

> Of all one colour with the blood a flowre she there did fynd . . .
> . . . Howbeet the use of them is short.
> For why the leaves* doo hang so looce through lightnesse in such sort,
> As that the windes that all things perce, with every little blast
> Doo shake them of and shed them so, as that they cannot last.
> (Golding's translation, 1567)

But *anemone* may have been a graecizing of the Aramaic *Na'aman*, 'darling', 'pleasantness', a by-name for Adonis. 'The Arabs still call the anemone "wounds of the Naaman"' (Frazer, *The Golden Bough*).

* i.e. petals

ANGELICA (*Angelica archangelica*, familiar for its bright green candied stems). XVI century. Medieval Latin (*herba*) *angelica*, 'angel's plant'; in accordance with a tradition that an angel had revealed its efficacy against such epidemic sicknesses as cholera and the plague. It was also called *archangelica* – a late German account saying that the angel was the archangel Raphael – and *radix Sancti Spiritus*, 'root of the Holy Ghost', source of the English name Holy Ghost and equivalent names in other languages. See *Archangel*.

ANISE (*Pimpinella anisum*, native of the Mediterranean region, the fruits of which are aniseed). XIV century. *P. anisum*, esteemed medicinally and as a flavouring and an aphrodisiac in the ancient world, was called *anneson* by the Greeks; Latin *anesum*, *anisum*, whence Old (and modern) French *anis*, and 14th century English *anys*, *anyse*.

ANISEED (see *Anise*, above). XIV century.

ANTIRRHINUM (species of the genus *Antirrhinum*, especially garden forms of *A. majus*, from S.E. Europe, introduced in the 16th century). XVI century. From *antirrhinum*, from Greek *antirrhinon*, *anti* + *rhin*–, 'like a nose or snout' (*rhis*), describing the snoutlike projection of the corolla, commonly likened to an animal's snout or mouth, as in such names for antirrhinum as *Calf's Snout*, *Weasel's Snout*.

APPLE (fruit of *Malus sylvestris*, Crab Apple, and orchard descendants of the anciently introduced *M. sylvestris ssp. mitis*, native of S.W. Asia and S.E. Europe). Old English *æppel*, from Germanic **aplu*, borrowed word of unknown origin ancestral to apple words in all Germanic languages. Celtic cognates (Welsh *afall*, Cornish and Breton *aval*, Gaulish *aballo-*) give the *Avalon* of Arthurian romance and *Avallon* in the Yonne department of France, both (like English *Appleton*) meaning 'apple orchard'.

APRICOT (*Prunus armeniaca* and fruit, fresh or dried). Mid XVI century. The name originates in the Latin *praecox*, 'early', 'early ripening'. The apricot (which the Romans first knew as *Armeniaca*, the Armenian tree, since they had it from Armenia, perhaps in the 1st century BC, after Pompey's Armenian campaign) was one of the *praecoces*, the early flowering and early ripening fruit-trees. Eventually it was called the *praecoqua*; and the fruit the *praecoquum*, which via Greek, Arabic (*al-barqūq*) and Portuguese (*albricoque*) gave the Tudor botanists and gardeners our older form, *apricock* (as in Shakespeare, *Richard 11*, iii, 4), ousted eventually by the parallel, and more easily spoken, *apricot*, from the French *abricot* (Miller's *Gardener's Dictionary* of the early 18th century, e.g. the edition of 1741, still lists the apricot under 'Apricock or Abricot').

ARBOR VITAE (species of the genus *Thuja*, especially *T. occidentalis*, *T. plicata*, and *T. orientalis*, the Chinese Arbor Vitae). XVII century, Evelyn 1664. Latin 'tree of life', given in 17th century to *T. occidentalis* from eastern North America on account of its evergreen, pungently scented foliage and undecaying timber (used, e.g., for the stockading of forts against Red Indians) and its inevitable association with the undecaying, perfumed, evergreen Cypress (q.v.) of Mediterranean grave-yards. *T. occidentalis* (more easily grown in the north than the half-hardy Cypress) soon entered English graveyards as a symbolic denizen.

ARBUTUS, ARBUTE (*Arbutus unedo*, Strawberry Tree). XVI century. From the Latin name *arbutus*. Often mentioned by Latin poets, e.g. Virgil, *Eclogues*, iii, 82: *dulce satis umor, depulsis arbutus haedis*, 'rain is sweet to the young corn, arbutus to the weaned kids.'

ARCHANGEL (various species of *Lamium* and allied genera, especially *Galeobdalon luteum*, Yellow Archangel). XVI century, Turner 1551. Translation of the medieval Latin *archangelica*, by which kinds of dead nettle were known as early as the 10th century. Since dead nettles resemble the nettle, but do not sting, it looks as if an unrecorded legend lies behind *archangelica*, the 'archangel's plant' – as if nettles were deprived of their sting, and indicated for their virtues, by one of the archangels (cf. *Angelica*).

ARROWHEAD (*Sagittaria sagittifolia*). XVI century, Gerard 1597. Translation of the 16th century botanists' Latin *sagittaria* (*herba*), from Latin *sagitta*, arrow, describing the shape of the leaves (and the anthers). Gerard writes of the 'large and long leaves, in shape like the signe Sagittarius, or rather like a bearded broad arrowe heade'.

ARROW-ROOT (*Maranta arundinacea*, native of tropical America, introduced 1732). XVII century. English equivalent for French *herbe aux flèches*, the tubers having been used to draw poison from arrow wounds. XIX century, starch from the tubers of *M. arundinacea*.

ARSESMART (*Polygonum hydropiper*, Water-pepper). XVI century. 'Because if it touch the taile or other bare skinne, it maketh it smart, as often it doth, being laid into the bed greene to kill fleas,' Minsheu, *Ductor in Linguas*, 1617. *P. hydropiper* has variously been called *Culrage*, 14th century (French *cul*, 'arse' + *rage*, 'fury'); *Smerthole*, 15th century; and *Smartweed*.

ARTICHOKE (*Cynara scolymus*, native of Mediterranean Europe, introduced c. 1548). XVI century, *archecokks*, *artochockes*, *archichokes*. From Italian *articiocco*, from Spanish *alcarchofa*, from the Arabic name for *C. scolymus*, *al-kharsūf* (*al*, the definite article). See also *Jerusalem Artichoke*.

ARUM LILY (*Zantedeschia aethiopica*). XIX century (when *Z. aethiopica*, introduced from South Africa in 1731, was also known as White Arum, Trumpet Lily, and Lily of the Nile). *Z. aethiopica* is one of the *Araceae* (not a lily), plants with a typical spathe, such as *Arum maculatum*, Lords and Ladies, and *Dracunculus vulgaris* (*Arum dracunculus*), Dragon Arum; the latter was known to the Romans and Greeks as *aron*.

ASAFOETIDA, ASSAFOETIDA (medicinal and flavouring gum-resin from *Ferula foetida*, native of S.W. Asia). XIV century. *Asa-*, medieval Latin from Latin *laser*, with loss of the initial 'l', from

3. Artichoke (Besler, *Hortus Eystettensis*, 1613)

laserpicium, laser, the asafoetida plant, ultimately from Accadian *lasirbitu*, asafoetida; + the feminine of Latin *foetidus*, 'stinking'.

ASARABACCA (*Asarum europaeum*). XVI century, Turner 1568. In the apothecaries' medieval Latin variously known as *asarum*, from Greek *asaron*, the name which Dioscorides used for *A. europaeum*, and *baccara*, from the Latin *baccar*, *baccaris* (Greek *baccharis*), which seems to have been a cyclamen. In English it was also known as *Bacchar* and *Asarum bacchar*.

ASH (*Fraxinus excelsior*). Old English *æsc*, with cognates in many European languages, from an inferred Indoeuropean base *os, osis.* A very ancient name, as might be expected for a tree whose timber was of major importance in prehistoric economy.

ASHWEED (*Aegopodium podagraria*). XVI century, Lyte 1578. From its persistence as a weed and the resemblance of its leaves to those of the ash (so also *Ground Ash*). Cf. *Ground Elder*.

ASPARAGUS (*Asparagus officinalis* ssp. *officinalis*, the plant and the young stems as vegetable). XVI century. From Latin *asparagus*, Greek *asparagos*, ? from a root meaning to sprout. Asparagus was cultivated in the Ancient World.

ASPEN (*Populus tremula*). Old English *æspe*, with cognates in other Germanic languages, replaced from the 16th century by its adjective *aspen* (cf. *oaken, ashen, elmen*: 'In English Aspe, and Aspen tree', Gerard 1597), though dialects have retained *asp* and *aps*.

ASPHODEL (the flower of the Elysian Fields, the 'asphodel meadow' inhabited by the souls of the dead, according to Homer; so species of the genus *Asphodelus* and the Yellow Asphodel, *Asphodeline lutea*, of the Mediterranean region; also Bog Asphodel, *Narthecium ossifragum*. Both the yellow *Asphodeline lutea* and species of *Asphodelus*, which are white-flowered, seem to have been called asphodel by the Ancients). XVI century, Gerard 1597. From Latin *asphodilus*, Greek (a loan-word) *asphodilos*. As well as asphodel, Tudor botanists still used the

4. Asphodel (Fuchs, *De Historia Stirpium*, 1542)

older *affodil* (14th century), from the medieval Latin *affodillus* (for *asphodelus*). See *Daffodil*.

ASPIDISTRA (*Aspidistra elatior*, introduced from China 1822). XVIII century. English use of the botanists' Latin, from Greek *aspis*, '(round) shield' + *astron*, 'star', in reference to the mushroom-like stigma.

ASTER (species of the genus *Aster*; also the China Aster *Callistephus chinensis*, and its many garden forms). Latin *aster*, from Latin *aster atticus*, Greek *aster attikos* (*aster*, star, + *attikos*, Attic, of Attica), names of a plant having 'little heads rayed in the likeness of a star' (Pliny), taken to be *A. amellus*, the Italian Aster or European Michaelmas Daisy (described by Virgil in the *Georgics* as *amellus*). Aster became established in English only towards the end of the 18th century. When garden authorities back to the 16th century wrote 'aster' they were consciously using the Latin word, and as a rule they added the common English name *Starwort*, first recorded in the 16th century.

AUBERGINE (fruit of *Solanum melongena* var. *esculentum*, the Eggplant, native of Africa and Asia, introduced late in the 16th century). XIX century. From the French *aubergine* ('Aubergines or Brinjals, which are highly esteemed in France, and may occasionally be met with in Covent Garden Market', Lindley and Moore, *Treasury of Botany*, 1874 ed.), from the Catalan *alberginera*, from Arabic *al-bādinjān* (*al*, the definite article), from Persian *badin-gan*, from Sanskrit *vatin-ganah*.

The earlier English or Anglo-Indian names *Pallingenie* and *Berenjaw* (16th century), and *Brinjal* (17th century; West Indian *Brown-jolly*) also descend from the same Sanskrit word by way of Persian, Arabic and Portuguese. The Sanskrit *vatin-ganah* means (a vegetable of the) 'wind-disorder class', i.e. a vegetable good against farting.

AUBRIETIA (*Aubrieta deltoides*, native of the S.E. Mediterranean region, introduced 1710). XIX century. English use of the botanists' Latin name (properly *aubrieta*, not *aubrietia*) given to the genus in memory of the great French botanical artist Claude

Aubriet (1665–1742), who accompanied Tournefort on his Levantine travels 1700–1702, during the course of which *A. deltoides* was discovered.

AURICULA (*Primula auricula*, from the Alps, introduced towards the close of the 16th century). XVIII century (XVII century, *auriculus*). English adoption of the botanists' Latin *auricula* (*ursi*), 'bear's ear', from the roughness and shape of the leaves. Gerard 1597 calls *P. auricula* both Bear's Ear and Mountain Cowslip.

AVENS (*Geum urbanum*; also *Geum rivale*, Water Avens, and *Dryas octopetala*, Mountain Avens). XIII century *avence*, from Old French *avence*, from medieval Latin *avancia*. See *Herb Bennet*.

AVOCADO (fruit of the tropical American tree *Persea americana*; and the tree itself). XVII century. A garbled survival of *ahuacatl*, the Nahuatl (Aztec) name for the fruit, corrupted in Spanish to *avocado*; from which came the older English forms *avogato* and *avigato*, in turn garbled to *alligator* (*pear*). (The fruit was also known as *Subaltern's Butter* and *Midshipman's Butter*.)

AWLWORT (*Subularia aquatica*). XVIII century. Botanists' translation of *subularia* (Latin *subula*, awl), plant with awl-shaped leaves.

AZALEA (various species and horticultural hybrids of the genus *Rhododendron*). XVIII century. English adoption of the botanists' Latin *azalea*, from the Greek *azaleos*, 'arid', 'dry', i.e. plant growing in dry conditions. The name was given by Linnaeus in 1737, as *Azalea procumbens*, to the species now called *Loiseleuria procumbens*, known to Linnaeus from dry habitats in Lapland.

B

BACHELOR'S BUTTONS (many species with round, ball-like flowers, double flowers; or flowers aggregated into heads like cloth buttons, e.g. *Tanacetum vulgare, Knautia arvensis, Centaurea nigra*). XVI century. The name derives from forms of love divination by which button flowers were picked by girls, given names of bachelors, and then put under their aprons or shifts; the flower which opened first indicated the bachelor to marry or the one who wished to marry the girl.

BACON AND EGGS (*Lotus corniculatus*). XIX century. Descriptive of the red and yellow of the flowers. Cf. the dialectical German *Eierbloom*, 'egg flower'.

BALDMONEY (*Meum athamanticum*). XVI century (XIV century for species of gentian, used in medicine). From the medieval Latin name *baldemonia*.

BALM (*Melissa officinalis*, native of Central Europe, the Mediterranean region and Western Asia, introduced in the late Middle Ages). XV century. A plant, on account of its aromatic lemon scent, like a balm or balsam, from Latin *balsamum*, from Greek *balsamon*, Balm of Gilead, i.e. oil or resin of the balsamon tree (*Commiphora opobalsamum*, from Arabia), from Hebrew *bāsām*, 'scent', 'balsam'.

BALM OF GILEAD (as a plant now generally the hybrid *Populus gileadensis*, of unknown provenance, but North American parentage. Also for *Populus tacamahaca*, Balsam Poplar, which is probably one of the parents of *P. gileadensis*). XVIII century. From the balsam-like perfume of the leaf-buds and unfolding leaves, and *Jeremiah* viii, 22, 'Is there no balm in Gilead; is there no physician there?' See *Balm, Balsam Poplar*.

BALSAM (species of the genus *Impatiens*, belonging to the family *Balsaminaceae*). XVIII century. See *Balm*.

BALSAM POPLAR (*Populus tacamahaca*, from North America, introduced 1792). XIX century. See *Balm of Gilead*.

BAMBOO (species of the genus *Bambusa* and other woody grasses, from India and the Far East). XVI century, *bambus*. From Dutch *bamboes*, from Portuguese *mambu*, from Malay *mambu*. Bamboo (17th century *bambou, bamboe*) is a singular assumed for reference to a single cane and to bamboo as a substance.

BANANA (fruit, and tree, of species of the tropical genus *Musa*). XVI century. From Portuguese or Spanish *banana*, for the fruit (the tree is *banano*). The Portuguese brought back the name from the Congolese coast, in the 16th century. Its origin is either a West African word *banam* or Arabic *banana*, 'finger' (from the shape of the fruit).

BANEBERRY (*Actaea spicata*, Herb Christopher). XVIII century. *bane*, 'poison', + *berry*. Constructed on the analogy, e.g., of *henbane* (q.v.). The black glistening berries are poisonous.

BANKSIAN ROSE (*Rosa banksiae*, from West China, introduced from a Cantonese garden in 1807 by William Kerr). 1837. From the botanical Latin *Rosa banksiae*, given in honour of Lady Dorothea Banks, wife of Sir Joseph Banks (1743–1820), President of the Royal Society.

BANYAN (*Ficus benghalensis*, native of India). XVII century. From the Sanskrit word for merchant (Hindi *vānija*); a *vānija* or banian was one of a caste of Indian traders. The banians trading at Bandar 'Abbas (then Gombrun) on the Persian coast were observed by the traveller Sir Thomas Herbert in 1628 to have a little cloth-decorated temple under a large *F. benghalensis*. He and his friends knew it as the Bannyan Tree, and from the account he wrote of his Persian travels banian or banyan was adopted as the English name for this huge and peculiar species.

BAOBAB (*Adansonia digitata*, native of tropical Africa). XVII century. Recorded first in 1592, as *bahobab*, by the Paduan physician and botanist Prospero Alpini, in his *De Plantis Aegypti*.

BARBERRY (*Berberis vulgaris*). XV century. From medieval Latin *barbaris*, *berberis*. Of unknown origin (? connected with *barb*, in reference to the arrowlike, tripartite spines of *B. vulgaris*. In the Vosges a name for *B. vulgaris* is *barbelin*, cf. French *barbillon*, 'a barb').

BARLEY (*Hordeum vulgare*, and its grains). Old English *bærlic*, *bær*, *bere*, 'barley' + noun suffix *-lic*, as in other plant names such as *charlock*, *hemlock* (Old English *cerlic*, *hymlic*). Barley – or *bere* – words include barton, 'barley farm', the place-name Berwick, 'barley farm', and barn, a 'barley house'.

BARRENWORT (*Epimedium alpinum*, from Southern Europe). Coined by Gerard 1597: 'being drunke it is an enimie to conception'. It was taken to be the *epimedion* of ancient medicine, of which Pliny says, 'Women should beware of it. Its leaves bruised in wine inhibit the breasts of virgins.'

BARTLETT (variety of dessert pear). XIX century. The American name for the William pear, after Enoch Bartlett of Dorchester, Mass., who introduced it to the United States *c.* 1820. See *William, Bon Chrétien*.

BASIL (*Ocimum basilicum*, native of tropical Asia, and Africa, introduced in the 16th century). XV century. From Old French *basile*, Medieval Latin *basilicum*, Greek *basilikon* (*phuton*), 'kingly (herb)', for its extreme pungency and many consequent uses in cookery and medicine.

BATH ASPARAGUS (*Ornithogalum pyrenaicum*). XIX century. Asparagus + *Bath* (the city). *O. pyrenaicum*, uncommon in England except in a few counties, especially Somerset, Wiltshire and Gloucestershire, is sold in an 'asparagus state' in shops in Bath and Bristol (as in markets in France); and is variously known as Bath A., French A., Prussian A., and French Grass ('grass', once a common greengrocer's term for asparagus).

BAY (-tree) (*Laurus nobilis,* native of the Mediterranean region, introduced in the 16th century). XV century *baie,* a berry, from Old French *baie,* Latin *baca;* bay-tree is thus 'berry-tree', the tree producing the *baccae lauri* (*L. nobilis* has black egg-shaped berries) which were much prescribed in medicine.

BAYBERRY (*Myrica cerifera,* from North America, introduced 1699). XIX century. Earlier (Lyte 1578) the berry of the bay (*Laurus nobilis*).

BEAN (*Vicia faba,* the Broad Bean, native of North Africa and S.W. Asia, anciently introduced; other leguminous vegetables; seeds of *V. faba,* etc., in and out of the pod). Old English *bēan,* with related names in most other Germanic languages.

BEARBERRY (*Arctostaphylos uva-ursi*). XVIII century. Translation of the botanists' Latin *arctostaphylus,* from a Greek plant name *arktou staphyle,* 'bear's grapes'.

BEAR'S-BREECH (*Acanthus mollis,* from Southern Europe, introduced in the 16th century). XVI century. A name additional to Brank-ursine, which derives from the shape of the flower, 'bear's claw'. *Breech* should mean 'backside', 'rump', which is hardly suggested by the leaf or the flower. See *Acanthus, Brank-ursine.*

BEAUTY OF BATH (variety of summer dessert apple). XIX century. Apple raised at Ballbrook, Batheaston, and introduced by the firm of Cooling, of Bath, c. 1864.

BEDEGUAR (the briar ball or Robin's Pincushion produced by the gall wasp *Rhodites rosae,* on stems of wild rose). XVI century. From French *bédegar,* from Persian *badawar,* 'brought by the wind', properly a species of thistle (?), which was mentioned by Arab botanists and which Tudor authors assumed to be the Milk Thistle, *Silybum marianum.* Interpreted also as 'wind-rose', as if Persian *bād,* wind, + *ward,* 'rose', and used by Tudor apothecaries to describe the growths on the wild rose. These bedeguars or briar balls (and the grub inside) were prescribed medicinally.

BEDSTRAW. See *Lady's Bedstraw*.

BEE ORCHID (*Ophrys apifera*). XVI century, Gerard 1597, 'Humble Bee Orchis'. Cf. Lyte 1578, orchis 'which hath in his floure a certayne figure of a Doure, or Drone Bee'. See *Orchid*.

BEECH (*Fagus sylvaticus*). Old English *bēce*, with congruent names in other Indoeuropean languages, including Latin *fagus*. Possibly for a tree with edible fruit, in this case the mast of the beech. In Greek *phegos* is oak, not beech, in fact the Valonia Oak, *Quercus macrolepis*, with large acorns which were sometimes eaten by man (this is the oak of the oracle of Zeus at Dodona).

BEECH FERN (*Thelypteris phegopteris*). XIX century. A confusing if deliberate and understandable mistranslation of the 'trivial' name *phegopteris*, Greek *phegos* (Valonia) oak + *pteris* fern, which Linnaeus coined for this species in his genus *Polypodium*. His genus also included the so-called Oak Fern (which does not grow on oaks, any more than the Beech Fern grows on beeches), for which he borrowed *dryopteris*, the Greek name (*drus*, 'oak', + *pteris*) for a fern which actually grew on oak trees. Since there could not be two Oak Ferns in the same genus, the *phego-pteris* was Englished as Beech Fern, as if *phegos* was the same as Latin *fagus*, 'beech'.

BEEFING, BIFFIN, BEAUFIN (variety of cooking apple from Norfolk). XVIII century. From *beef*, describing the bright red colour of the fruit.

BEER (drink now made from malted barley flavoured with hops). Old English *bēor*, from monastic Latin *biber*, 'a drink'; from Latin *bibere*, 'to drink'.

BEET (*Beta vulgaris*, including beetroot, sugar-beet, etc., and the wild beet). From Latin *beta*, borrowed early in Germanic languages. In Old English *bēte*, a word apparently lost – as if beet was little cultivated – and reintroduced late in the 14th century from the Low German of North Germany. By the Ancients beet was grown as a leaf-vegetable (cf. our spinach-beet, from the white form).

BEGONIA (horticultural forms of *Begonia semperflorens*, native of Brazil, introduced in 1829). Named after Michel Bègon (1638–1710), governor of the French colony of St Dominique, in the island of Haiti. (The earliest begonias to be cultivated in Europe came from the West Indies.)

BELL-FLOWER (species of the genus *Campanula*). XVI century. Descriptive of the bell-like corolla (the medieval Latin was *campanula*, 'little bell', diminutive of Latin *campana*, 'a bell').

BENT (various kinds of stiff, coarse grass). Old English *beonet*, which is found in place-names, such as Bentley, Bensted, Bentham.

BERGAMOT, WILD BERGAMOT (*Monarda fistulosa*, introduced 1686; and *M. didyma*, Oswego Tea, introduced 1752. Both from North America). A Turkish pear, the *beg armydu*, 'prince's pear', was called in Italian *bergamotta*, French *bergamotte*, English (17th century) *bargamot*. The name was also given to *Citrus bergamia*, grown in Italy for its essential oil, essence of bergamot; and so to the Lemon Mint *Mentha citrata*, and (19th century) to the aromatic *Monarda fistulosa* of North America. See *Oswego Tea*.

BERGAMOT (variety of pear). See above.

BERRY (roundish fruit or drupe). Old English *berige*, *berie*, with cognate words in other Germanic languages.

BERMUDA GRASS (*Cynodon dactylon*). XIX century (U.S.A.; where it is planted as a lawn grass). From its occurrence in Bermuda.

BETEL NUT (fruit of the palm *Areca cathecu*, from India). XVII century. From Portuguese *betel*, from Malayalam *vettila*, the plant *Piper betle*, leaves of which are chewed with lime and parings of Areca nut. 'The leaf is the delight of the Asiatics; for men and women, from the prince to the peasant, have no greater pleasure than to chew it all day in company; and no visit begins or ends without this herb. – The *betel* makes the lips so fine, red, and beautiful, that if the European ladies could they would purchase

23

it for the weight in gold' (Vieyra's *Dictionary of the Portuguese and English Languages*, 1813 ed.).

BETONY (*Stachys officinalis*). XIV century. From Old French *betoine*, from medieval Latin *betonica*, Latin *vettonica*. 'The Vettones in Spain discovered (i.e. as a medicinal herb) the plant known in Gaul as vettonica . . . which is reputed above all others' (Pliny).

BHANG (*Cannabis sativa*, especially for smoking and chewing). XVI century *bangue*, from Portuguese *bangue*; XVII century *bang*. From Hindi *bhāng*, from Sanskrit *bhångā*.

BIGARREAU (firm-fleshed varieties of cherry). XVII century. French *bigarreau*, from *bigarré*, 'variegated' (in colour, generally cherries which are both red and yellow), from Old French *garre*, 'variegated', a word of Germanic origin.

BIG TREE (*Sequoiadendron giganteum*, Wellingtonia; from California, introduced 1853). XIX century. *S. giganteum* is the bulkiest tree in the world. See *Wellingtonia*.

BILBERRY (*Vaccinium myrtillus*, the plant and fruit). XVI century. The first element is probably a Scandinavian name; of unknown meaning. See *Blaeberry*.

BILDERS (*Apium nodiflorum*; and *Heracleum sphondylium*, Hogweed). Old English **billere*, which occurs in several place-names, such as Bilbrook, Somerset, earlier *bilrebroc*, in which it seems most likely to refer to *A. nodiflorum*. Cf. the Brunswick names for *Berula erecta* (which resembles *A. nodiflorum* in leaf and habitat), *beckbille* and *beekbilder*, 'brook' + *bille*, *bilder*, referable to German *berle*, from Latin *berula*, of Celtic origin, cf. Irish *biolar*, Cornish and Breton *beler*, 'water-cress'.

BINDWEED (*Calystegia sepium* and *Convolvulus arvensis*). XVI century. *bind* + *weed*. The earlier name was the still common *withwind*, q.v.

BIRCH (trees of the genus *Betula*, especially the European White Birch, *B. verrucosa*). Old English *beorc*, *birce*. German *birke*, etc.

One of the ancient Indoeuropean tree names, from a root meaning 'white' or 'shining'.

BIRD-CHERRY (*Prunus padus*). XVI century, Gerard 1597, 'The Birds Cherry-tree'.

BIRD'S EYE (*Primula farinosa*, Bird's Eye Primrose). XVI century, Gerard 1597, *Birds Eine*, from the yellow throat of the corolla tube in the centre of the flower.

BIRD'S FOOT (*Ornithopus perpusillus*). XVI century, Lyte 1578, translating the Dutch *vogelvoet*, or the 16th century botanists' Latin *ornithopodium* (Greek *ornithos*, genitive of *ornis*, 'bird', + *podion*, 'little foot'), describing the curved pods.

BIRD'S FOOT TREFOIL (*Lotus corniculatus*). XIX century. From its resemblance to Bird's Foot (*Ornithopus perpusillus*), from which it is distinguished by the trifoliate leaves.

BIRD'S NEST (the orchid *Neottia nidus-avis*). XVI century, Lyte 1578, translating the German *Vogelnest*, which itself translated the Latin name *nidus avis* given to the plant in the 16th century, 'bicause that the rootes be so tangled and wrapped like to a nest' (Lyte). The name (Gerard 1597) was also given to the Wild Carrot (*Daucus carota*), the umbel of which in seeding draws together into a bird's nest shape.

BIRTHWORT (*Aristolochia clematitis*). XVI century, coined by William Turner as a translation of its medieval Latin name *aristolocia*, from Greek *aristolochia*, *aristos*, 'best', + *locheia*, 'childbirth', name of a species anciently given to assist birth (Theophrastus, Dioscorides, Pliny).

BISHOP'S WEED (*Aegopodium podagraria*, Ground Elder, Goutweed, Herb Gerard). XIX century, Anne Pratt 1861. A name which had been applied in the 17th century to the medicinal plant *Ammi majus*, from southern Europe.

BISTORT (*Polygonum bistorta*). XVI century. From the medieval Latin name *bistorta*, 'twice twisted', in reference to the contorted rhizome.

25

BITTER-SWEET (*Solanum dulcamara*, Woody Nightshade). XVI
century, William Turner's equivalent of the medieval Latin
names *dulcamarum, dulcis amara, amara dulcis* (*dulcis* sweet, *amarus*
bitter), in reference to the taste of the berries – 'faire berries . . .
very red when they be ripe, of a swete taste at the first, but after
very unpleasant, of a strong savour; growing togither in clusters
liked burnished corall' (Gerard).

BLACKBERRY (*Rubus fruticosus*, the fruit and the plant). Old
English *blaceberian* (plural).

BLACK BRYONY (*Tamus communis*). XIX century. Translation
of Latin *bryonia nigra*, the *vitis nigra* or Greek *ampelos melaina*,
'black vine', of ancient medicine. See *Bryony*.

BLACK-EYED SUSAN (*Thunbergia alata*, native of tropical
Africa, introduced 1823; also *Rudbeckia hirta*, from North
America, introduced 1714). For *T. alata*, the dark throat of the
flower giving it the appearance of an eye, XIX century. For the
black centred *R. hirta*, XX century. One of the large number, old
and modern, of Christian names affectionately applied to a
flowering plant.

BLACK PEPPER. See *Pepper*.

BLACKTHORN (*Prunus spinosa*). Old English **blæc-thorn*. XIV
century *blakthorn*.

BLADDER CAMPION (*Silene cucubalus*). XIX century. Descriptive of the swollen calyx. See *Campion*.

BLADDERWORT (species of the genus *Utricularia*). XIX century.
An English equivalent of the generic name *Utricularia*, from
utriculus, 'a small leather bottle', in reference to the plant's
insect-catching bladders.

BLADDER WRACK. See *Wrack*.

BLAEBERRY (*Vaccinium myrtillus*, Bilberry, Whortleberry; and
the fruit). XV century. *blae*, 'blue-black', + *berry*. In Old Norse
blaber, Norwegian *blabeer*. See *Bilberry*.

BLEEDING HEART (*Dicentra spectabilis*, native of Japan, introduced by way of China 1846). XIX century. Describing the rosy colour and heart shape of the pendulous flowers. Cf. German *Flammendes Herz*. See *Dutchman's Breeches*.

BLENHEIM ORANGE (variety of apple). XIX century. Raised from a pip by a Mr Kempster at Woodstock, Oxfordshire, and named for its yellow colour and after the Blenheim Palace of the Dukes of Marlborough alongside the town. The tree was becoming known c. 1818.

BLESSED THISTLE (*Silybum marianum*, Lady's Thistle). XIX century, but properly (Lyte 1578) the name of *Cnicus benedictus* of Southern Europe, which was much used in medicine, and was therefore the *carduus sanctus* or *carduus benedictus*, the 'sacred' or 'blessed thistle' (Latin *benedictus*, 'blessed'). See *Milk-thistle*.

BLEWIT (the edible fungus *Trichloma saevum*). XIX century. From the violet-blue stipe. Cf. *bluet*, French *bleuet*, for *Centaurea cyanus*. The name is wrongly given in many books, following the *Oxford English Dictionary*, as blewetts, which is simply the plural. See *Blue Leg*.

BLIND-NETTLE (*Lamium album*, and related nettle-like species without stinging hairs). Old English *blind netel*; also Deaf-nettle, as if, either way, the lack of stinging hairs could be described only in the loss of an animal sense. See *Dead-nettle, Nettle*.

BLINKS (species of the genus *Montia*). XIX century. ? from the description of *M. verna* (*M. fontana*) in the Latin of Christopher Merret's *Pinax Rerum Naturalium Britannicarum*, 1667, as the *alsine* 'with little blinking flowers'.

BLUEBELL (generally in England *Endymion nonscriptus*. In northern counties and Scotland *Campanula rotundiflora*, Harebell). XVI century for *Campanula* species (Lyte 1578), cf. the Norwegian *blaaklokka* for *C. rotundiflora*. There is no record of bluebell for *E. nonscriptus* before 1794. In Gerard 1597 *E. nonscriptus* is *Hyacinthus Anglicus*, Blew English Hare-bels, English Jacint; in Turner 1548 'crowtoes, and in the North partes Crawtees'.

BLUE LEG (the edible fungus *Trichloma saevum*). From the violet-blue of the stipe. Cf. the French *pied bleu*. See *Blewit*.

BLUSHER (the edible fungus *Amanita rubescens*). XIX century. Equivalent of the Latin *rubescens*, 'blushing', in the specific name, the flesh of this fungus turning red when broken or bitten.

BOG ASPHODEL (*Narthecium ossifragum*). XIX century. A translation of the 16th century botanists' Latin *asphodelus* (*luteus*) *palustris*, as if the plant were a yellow asphodel of the garden (*Asphodeline lutea*) of miniature size.

BOGBEAN. See *Buckbean*.

BOG MYRTLE (*Myrica gale*, Sweet Gale, Gale). XIX century, though Gerard 1597 had used the names *Dutch Myrtle tree* and *Wilde Myrtle*. *Myrtle tree* translated the German *Mirtelbaum*. In Germany the fragrant *M. gale* had long been taken as a kind of *Myrtle* (*Myrtus communis*) – Dutch Myrtle tree standing for the botanists' Latin *myrtus brabantica*, 'myrtle of Brabant'.

BOLETUS (fungi of the genus *Boletus*, especially *B. edulis*). XIX century. English use of the Latin *boletus* (name of the fungus most commonly eaten in the Ancient World), from the Greek *bolites*, from *bolos*, 'clod' or 'lump', from the shape of a Boletus.

BON CHRÉTIEN (varieties of dessert pear). XVI century. French 'good Christian', from the Latin (*poma*) *panchresta*, 'all-good (fruit)', assimilated to *bon Chrétien*. The pear was brought from Italy to France, to Plessis-les-Tours, by St Francis of Paolo, who came to Plessis in 1482 to attend the sick Louis XI. The altered name is thought to have been devised by the king. See *William*.

BORAGE (*Borago officinalis*, native of the Mediterranean region. Introduced in the Middle Ages). XIII century. Old French *bourrache*, from medieval Latin *bor*(*r*)*ago*, a name derived from Arabic *abū 'ārak*, 'father of sweat'. Borage was long given to produce sweat and drive out fever.

BORECOLE (cultivated forms of *Brassica oleracea*). XVIII century, when the common spelling was *boor-cole*; from Dutch *boerenkool*, 'boors' kale', 'peasants' kale'. See also *Cole, Kale*.

BO-TREE (*Ficus religiosa*, native of India). XIX century. 'Tree of Enlightenment': Sinhalese *bogaha*, *bo*, 'enlightenment' or 'perfect knowledge', + *gaha*, 'tree'; from Pali *bodhitarū*. Under a *F. religiosa* Gautama became *buddha*, the enlightened.

BOUGAINVILLEA (*Bougainvillea spectabilis* and *B. glabra*, natives of Brazil). XIX century. English use of the botanists' Latin name which was given in honour of the French naval officer, circumnavigator and man of science, Louis Antoine de Bougainville (1729-1811).

BOUNCING BETT (*Saponaria officinalis*, Soapwort; *Kentranthus ruber*, Red Valerian, native of Central Europe and the Mediterranean region, introduced in the 16th century). XIX century. A suitable name for *S. officinalis*, which is irrepressible in a garden, and *K. ruber*, bouncingly taking possession of old walls.

BOURBON ROSE (*Rosa × borboniana*). 1829. English equivalent for the French *Rosier de l'Isle Bourbon*, the original Bourbon Rose having been introduced into France, in 1819, from Ile de Bourbon (Réunion) in the Pacific.

BOURTREE (*Sambucus nigra*, Elder). XV century. A name chiefly of north-western counties and Scotland. ? from Old Norse *býjar*, genitive singular of *býr*, + *tré*, 'village tree', 'farmstead tree'. The elder is commonly found in neglected corners of villages and farmsteads.

BOX (*Buxus sempervirens*). Old English *box*, from Latin *buxus*, from Greek *puxos*. A box was originally a receptacle made of boxwood.

BOYSENBERRY (form of *Rubus logano-baccus*, the Loganberry; and its fruit). XX century. From the name of the American horticulturalist Rudolph Boysen, *floruit* 1923, who raised it.

BRACKEN (*Pteridium aquilinum*). Old English **brǽcen*, ? *bracu*, 'brake', 'bracken', + the substantival collective suffix *-en*; or from Old Norse **brakni*, which is found in North Country place-names. See *Brake*.

BRAKE (*Pteridium aquilinum*). Old English *bracu*. Possibly in the sense of something broken and cognate with the verb to *break*, which would be appropriate to the dead fern, forming a characteristic tangle of broken stems.

BRAMBLE (*Rubus fruticosus*). Old English *brǽmbel*, from *brom*, wiry or thorny shrub (as in *broom*), + the substantival concrete suffix *-el*.

BRAMLEY'S SEEDLING (variety of cooking apple). XIX century. Apple raised at Southwell, Nottinghamshire, by Matthew Bramley, butcher and innkeeper, and introduced commercially in 1876.

BRANDY (alcoholic drink distilled from wine). XVII century, first as *brandewine*, *brandwine*, shortened in second half of the century to *brandee*, *brandy*. From Dutch *brandewijn*, 'distilled wine' (*branden*, to distil, literally to burn).

BRANDY-BOTTLE (*Nuphar lutea*, Yellow water-lily). XIX century (1846). Descriptive of the scent of stale alcohol in the flowers and the flask- or bottle-shaped fruit.

BRANK-URSINE (*Acanthus mollis*, from Southern Europe, introduced in the 16th century). XV century, *brankurcyna*. From the medieval Latin *branca ursina*, 'bear's claw', describing the shape of the flower arched by the upper lip of the corolla. See also *Acanthus*, *Bear's-breech*.

BRAZIL-NUT (nut from the fruit of species of *Bertholletia*, especially *B. excelsa*, huge tropical forest trees of Brazil, Guiana and Venezuela). XIX century.

BREAD (milled cereal grain baked with yeast). Old English *brēad*, with related words in most other Germanic languages, from an Indoeuropean base meaning to ferment, ancestral also to the word *brew* and *broth*.

BRIAR (roots and stems of *Erica arborea*, Tree Heath, native of the Mediterranean region, imported since mid 19th century for making briar pipes). XIX century, at first as *bruyer, bruyer wood*. From French *bruyère*, heath, heather, from Low Latin **brucaria*, from *brucus*, a Low Latin word of Gaulish origin.

BRIAR, BRIER (the various species of Wild Rose). Old English *brēr, brǣr*, originally meaning a prickly bush, especially brambles. Since the 16th century more commonly a wild rose (from the contrast in poetry of the gentle rose and the sharp brier which produces it), though the old sense has not disappeared.

BROAD BEAN (*Vicia faba*, from North Africa and S.W. Asia, anciently introduced). XVIII century (1783). *Bean* (q.v.) was the word for *V. faba*. When other bean-like species were introduced from the New World in the 16th and 17th centuries, this ancient bean of the European past required a distinctive epithet.

BROCCOLI (*Brassica oleracea* var. *botrytis*). XVII century. Plural of Italian *broccolo*, North Italian word for Cauliflower, diminutive of *brocco*, 'shoot' or 'sprout'.

BROME-GRASS, BROME (grasses of the genus *Bromus*). XVIII century. Englished (from Latin *bromos*, oats) by the dilettante and botanist Benjamin Stillingfleet (1702–1771), who devised English names for various grasses.

BROMPTON STOCK (biennial forms of *Matthiola incana*). XVIII century (Miller 1731). The forms originated in the once famous Brompton Park Nursery, near London. See *Stock*.

BROOKLIME (*Veronica beccabunga*). Old English *hleomoce*, a plant name with the diminutive suffix *-oc*. XV century *brok-lemok*, 'brook', + the name. Brooklime (Old Norse *bekkrbung*, 'brook bung', German *bachbunge*) was in northern use as a salad and in

31

medicine. The Old English *hleomoce* occurs in the Northumberland parish name Lemmington.

BROOKWEED (*Samolus valerandi*). XIX century, Anne Pratt 1861. A local plant, for which the name was probably invented by Anne Pratt by analogy with *Brooklime*.

BROOM (*Sarothamnus scoparius*). Old English *brōm*, cognate with words in other Germanic languages meaning 'broom', 'bramble', etc. The base seems to have signified a thorny or wiry shrub.

BROOMRAPE (species of the genus *Orobanche*, especially *O. rapum-genistae*). XVI century. English equivalent of the botanists' Latin *rapum genistae*, 'root-knob of the broom', on which *O. rapum-genistae* is parasitic.

BRUSSELS SPROUTS (leaf-buds, and the plants, of *Brassica oleracea* var. *gemmifera*). Late XVIII century. 'Cultivated around Brussels from time immemorial; although it is only within the last twenty years it has become generally known in this country' (Lindley and Moore, *Treasury of Botany*, 1866).

BRYONY (*Bryonia dioica*). XVI century. From the medieval Latin *brionia*. *B. dioica* was taken to be the Latin *bryonia* or *vitis alba*, 'white vine'; Greek *bruonia* or *ampelos leuke*, of ancient medicine; from the Greek *bruein*, 'to grow luxuriantly'. See *Black Bryony*.

BUCKBEAN (*Menyanthes trifoliata*). XVI century, Lyte 1578. Translation of Flemish *bocksboonen*, which Lyte took to mean 'goat's beans'. The leaves resemble the young leaves of the Broad Bean, and *bocksboonen*, German *Bocksbohnen*, may be one of those names like horse-chestnut for a plant or fruit suited only to an inferior taste. It has also been suggested that the name is short for German *Scharbocksbohnen*, Scurvy Beans, since in German, Swedish, Danish and Flemish *M. trifoliata* has also been called by names equivalent to Scurvy Clover (German *Scharbocksklee*). Bogbean (late 18th century) is likely to be a transformation of buckbean to fit its habitat, as a more comprehensible name.

5. Bryony (Fuchs, *De Historia Stirpium*, 1542)

BUCKEYE (*Aesculus glabra*, native of the United States, introduced in 1812, and other species of *Aesculus*, Horse-Chestnut). Late XVIII century (America) from the resemblance of the shining brown fruit with its pale scar or hilum to the eye of a buck.

BUCKLER-FERN (ferns formerly grouped in the genus *Aspidium*). Buckler, for 19th century botanists' Latin *aspidium*, from Greek *aspidion*, 'small shield', in reference to the shape of the indusium or spore cover.

BUCKTHORN (*Rhamnus cathartica*). XVI century, Lyte 1578. Translation of *cervi spina*, 'buck's thorn', the early botanists' Latin for *R. cathartica*. See also *Sea-buckthorn*.

BUCKWHEAT (*Fagopyrum esculentum*). XVI century. Clever botanists in the 16th century noticed the resemblance between the dark-brown three-edged seeds and beech-mast; and assumed that the Low German names *boeckueyt*, *buchweiss* meant 'beech wheat' (whence the specific name *fagopyrum*, made up of Latin *fagus*, 'beech', and Greek *puros*, 'wheat'). But it seems likelier that the name (15th century *bukweten* in Low German) means 'goat wheat', for a grain inferior to true wheat.

BUDDLEIA (species of the genus *Buddleja*, especially *B. davidii*, native of China, introduced in 1896). Late XIX century. The generic *Buddleja* was invented by Linnaeus in honour of the English botanist, the Rev. Adam Buddle (d. 1715). As an English name Buddleia became general with the popularity of *B. davidii*. See *Butterfly Bush*.

BUGLE (*Ajuga reptans*). XIII century. From the Low Latin name *bugula* by which the plant was known to physicians and apothecaries.

BUGLOSS (*Lycopsis arvensis*). XVI century, in this sense. From the Greek plant-name *bouglossos*, 'ox-tongued', via Latin *buglossa*, and French *buglosse*. Bugloss was the name given to several plants with rough tongue-like leaves, the Greek *bouglossos* having been an *Anchusa*. See *Viper's Bugloss*.

6. Bugloss (Matthiolus, *Commentarii*, 1565)

BULLACE (*Prunus spinosa* ssp. *institia*; and fruit). XIV century. From the Old French *buloce*, 'a sloe'.

BULRUSH (*Typha latifolia*; in botanical literature *Scirpus palustris*). XV century. *bull* + *rush*, i.e. a rush of exceptional size and prominence. Cf. *bullfinch*.

BURDOCK (*Arctium lappa*). XVI century, Gerard 1597. Usually explained simply as *bur* + *dock*, i.e. a large coarse-leaved (dock-like) plant with adhering fruitheads. In the late Middle Ages the common name for *A. lappa* was Bur, cf. *bor*, *burre* in N.W. Germany. The *dock* of Gerard's *Burre Docke* may have had the original sense of ball. Cf. Frisian *dok*, 'ball', 'bundle', and N.W. German names for *A. lappa* mentioned below under *Hardokes*.

BUR MARIGOLD (*Bidens cernuus*, and *B. tripartitus*). XIX century. The achenes having barbed bristles which make the seeds stick, and the flower resembling a marigold.

BURNET (*1. Sanguisorba officinalis*, Great Burnet. *2. Poterium sanguisorba*, Salad Burnet. *3. Pimpinella saxifraga*, Burnet Saxifrage). In the first and second sense, XV century; in the third sense, XVI century. The flower-heads of Great Burnet and Salad Burnet are a dark crimson-brown or mahogany, or burnet (Old French *burnete*, *brunet*). Burnet Saxifrage was taken as a Burnet because the pinnate root-leaves resemble the leaves of Great Burnet and Salad Burnet. See *Pimpernel*.

BURNET ROSE (*Rosa spinosissima*). XIX century (Miller 1731, 'wild burnet-leaved Rose'). Rose with leaves like the Burnet (*Pimpinella saxifraga*), the early name for *Rosa spinosissima* having been *Rosa spinosissima foliis pimpinellae*, 'with leaves of the Pimpinella'.

BURNET SAXIFRAGE (*Pimpinella saxifraga*). XVII century. See *Burnet*, *Pimpernel* and *Saxifrage*.

BUTCHER'S BROOM (*Ruscus aculeatus*). XVI century. *R. aculeatus* was apparently used by butchers to wipe off their stalls, chopping blocks etc., and as a fly-whisk, though there is no detailed or convincing statement of this. Victorian butchers also decorated Christmas sirloins with the red-berried stems.

BUTTERBUR (*Petasites hybridus*). XVI century. *butter + bur*. Several large-leaved plants have *butter–* in their names in English and German, etc., as if from wrapping butter. *P. hybridus* is (Butter)bur – it has no burs – from its broad resemblance to *Arctium lappa*, Burdock, names for which include English *clot-bur*, North Frisian *bor*, Low German *bor, burre*.

BUTTER-BEAN (*Phaseolus lunatus*, native of tropical America, and its beans). Late XIX century, from the butter-yellow colour of the dried and cooked beans.

BUTTERCUP (meadow species of *Ranunculus*). Not recorded until late in XVIII century. Earlier names are Crowfoot, 15th century; Butterflower (cf. German *Butterblume*), Goldcup, Kingcup, Goldknop, 16th century; Giltcup, 17th century. *butter*, from the fancied connection between the yellow of Buttercups and the yellow of cream and butter from the cow's milk and *cup*, from the flower shape.

BUTTERFLY-BUSH (*Buddleja davidii*, native of China, introduced 1896, and other *Buddleja* species). XX century. Cf. German *Schmetterlingsstrauch*. The flowers of *B. davidii* attract butterflies to a remarkable degree. See *Buddleia*.

BUTTERFLY ORCHID (*Platanthera chlorantha* and *P. bifolia*). XVIII century ('Butterfly orchis'), though see Gerard 1597, 'that kind which resembles the white Butter-flie'.

BUTTERWORT (*Pinguicula vulgaris*). XVI century, Gerard 1597. Plant which encourages or protects the milk-producing capability of cows, so ensuring a supply of butter. It protected cows from elf-arrows; and human beings from witches and fairies. Gerard says farmers' wives in Yorkshire anointed cows' udders when bitten, chapped, etc., with its 'fat and oilous juice'. Cf.

Buxbaum's Speedwell

16th century (and modern) German *Butterwurz*, 'butter root', and *Fettkraut*, 'fat plant'.

Buxbaum's Speedwell (*Veronica persica*, from Western Asia, found growing in England in 1825). First described and pictured in 1727 by the German botanist Johann Christian Buxbaum (1683–1730), from fields outside Constantinople.

C

CABBAGE (*Brassica oleracea* var. *capitata*). XIV century *caboche*, from the Old French (of N.E. France, the area of the Somme) *caboche*, head. Other medieval names for this Brassica with a man-sized head were *frenchwoort* and *Caulus gallica*.

CABBAGE LETTUCE (*Lactuca sativa* var. *capitata*). XVI century, Turner 1562: 'Cabbage lettes, because it goeth all into one heade, as cabbage cole dothe.' See *Lettuce*.

CABBAGE ROSE (*Rosa* × *centifolia*). XVIII century. Descriptive of the double flowers, like heads of cabbage.

CACTUS (plants of the family *Cactaceae*). XVIII century. Linnaeus borrowed this Latin and Greek name for the cardoon (which has a prickly involucre) to describe the unfamiliar *Cactaceae* from the New World.

CALABASH (1. dry fruits or gourds of *Lagenaria siceraria*, native of the Old World tropics; 2. gourd-like fruits of *Crescentia cujete*, the Calabash-tree, native of tropical America). XVI century. In English, according to the *Oxford English Dictionary*, first on record in the second sense in Ralegh's *Discoverie of Guiana*, 1596. In the first, and usual, sense, XVII century (Evelyn 1658).

CALAMONDIN (the small acid orange of *citrus mitis*, the Cala-mondin Orange or Panama Orange; and the tree, native of the Philippines). XX century. From *Kalamunding*, the name in Taga-log, the national language of the Philippines.

CALCEOLARIA (species of the genus *Fagelia*, formerly *Calceolaria*, from South America, mostly introduced in the early 19th century). XIX century. English use of the botanists' Latin of 1725, from *calceolus*, 'slipper', + botanical suffix *-aria*, in reference to the shoe shape of the lower lip of the flower. The early 19th

39

century name Slipperwort has failed to establish itself, not without reason.

CAMELIA (species of the genus *Camellia*, from China and Japan, especially forms of *C. japonica* and *C. sasanqua*). XVIII century (as Camellia). English use of the Linnaean name honouring the Jesuit botanist Georg Joseph Kamel or Camellius (1661–1706), whose account of the plants of Luzon in the Philippines was published in 1704 by John Ray.

CAMPION (*Lychnis coronaria*, Rose Campion; *Melandrium rubrum*, Red Campion, *M. album*, White Campion). XVI century. This pretty name has not been convincingly explained. It is evidently an Elizabethan invention, first on record in 1576, and it was soon taken up by poets as well as botanists. Campion (= champion) was a word still understood; and the least implausible suggestion is that *Lychnis coronaria*, Rose Campion, was so called by someone given to poetic archaism, as if this tall and striking newcomer was a champion of the garden and the summer.

CANADIAN FLEABANE (*Erigeron canadensis*, native of North America). XX century. *E. canadensis* was described in the 17th century as *aster canadensis*, having been introduced from Canada in 1653 to a botanic garden at Blois in France, from which it soon spread through Europe, as a weed.

CANADIAN PONDWEED (*Elodea canadensis*, native of North America, first found as an invasive plant in Ireland in 1836). XX century. Describing its origin and habitat.

CANDYTUFT (*Iberis umbellata*, from Southern Europe, and garden forms). XVIII century. *Candy* (since *I. umbellata* was known in the 16th century as *thlaspi Creticum*, *thlaspi Candiae*, Cretan cress, cress of Candy, Englished as Candie Thlaspi, Candie Mustard), + *tuft*, from the corymbose habit of the plant.

CANNABIS (dried leaves – hashish, marijuana – of *Cannabis sativa*, Hemp, native of temperate Asia). XX century. English use of the generic name, Latin *cannabis*, from Greek *kannabis*; with which *hemp* is also cognate.

CANTALOUP (the warty melon *Cucumis melo* var. *cantalupensis*). XVIII century. From the French *cantaloup*, Italian *cantalupo*, cantaloups having been first raised, from Armenian seed, in the gardens of a papal villa at Cantalupo (Mandela), between Tivoli and Arsoli.

CANTERBURY BELLS (*Campanula medium*, native of Southern Europe). XVI century. A name (? originally for the wild *C. trachelium*) from the resemblance of the bell-shaped flowers to the latten St Thomas's Bells, sold as signs or badges to pilgrims to the shrine of St Thomas à Becket in Canterbury Cathedral (such bells of the 14th and 15th centuries may be seen, e.g., in the Guildhall Museum, London).

CAOUTCHOUC (solidified latex of the tree *Hevea brasiliensis*, native of the Amazon region). XVIII century. From the French *caoutchouc*, from Spanish *cauchuc*, a loan-word from the language of the Quechua Indians.

CAPE GOOSEBERRY (*Physalis peruviana*, from South America, berry and plant). Late XIX century. *P. peruviana* has been grown extensively in the Cape – the former Cape Colony, South Africa.

CAPER (pickled flower-bud of *Capparis spinosa*, native of the Mediterranean region). XV century, *capres*. From Old French *câpres*, from Latin *capparis*, from Greek *kapparis*. The *s*, taken to be a plural ending, was discarded in the 16th century.

CAPER SPURGE (*Euphorbia lathyrus*, Catapuce). XIX century. The fruits resemble capers.

CARAWAY (*Carum carvi*, native of Europe and temperate Asia, ? introduced in the 16th century). XV century. The name used by medieval doctors and apothecaries was *carvi*, from Arabic *karawiyā* (Latin *careum*, Greek *karon*). Caraway comes through Spanish *alcaravea* from the same Arabic *karawiyā*, plus the definite article *al*, and so from the Latin and the Greek. Pliny – but this was only folk etymology – derived *careum* from Caria, in S.W. Asia Minor, where he said caraway was grown.

CARDOON (*Cynara cardunculus*, native of Southern Europe, introduced in the 17th century). XVII century. From French *cardon* (16th century), from *carde*, 'prickly flower-head', Provençal and Italian *cardo*, 'thistle', from the Latin *cardus*. The name may be described as a transformation of the Latin *cardus*, *carduus*, 'thistle', in a special sense. *C. cardunculus* is a root vegetable, but like its relative *C. scolymus*, the Artichoke (q.v.), from which it may have been derived in cultivation, it bears a prickly thistle-like involucre. Roman authors, e.g. Pliny, use *carduus* not only in the ordinary sense of thistle, but as this special thistle or prickly-headed plant *C. cardunculus*. Cardoon in English was first applied to the artichoke.

CARLINE (Thistle) (*Carlina vulgaris*). XVI century, Lyte 1578, who repeats the statement by the French herbalist Jean Ruel in his *De historia stirpium* 1536 that this thistle was called *Carolina* because it was divinely revealed to Charlemagne as a remedy against the plague – 'bycause of Charlemaigne Emperour of the Romaynes, unto whom an Angel first shewed this Thistel, as they say, when his armie was striken with the pestilence.' From the French *Carlina*, from medieval Latin *Carolina* (*herba*) or *Carlina* (*herba*), from Latin *carduus*, 'thistle', altered by association with *Carolus*, i.e. Charles, on account of the legend.

CARNATION (*Dianthus caryophyllus*, the Clove Pink, in its various forms). XVI century. One of the new Tudor plant-names, with a strong poetic tone, which has not been satisfactorily explained. First (Turner 1538) as *Incarnacyon*; later *Carnation*, *Coronation*, *Cornation*. Elizabethan poets quickly accepted *Carnation*, which they employ as well as the usual *Gilliflower* (q.v.). There seems no reason not to accept William Turner's statement that *D. caryophyllus* was called *Incarnacyon* in common speech; which may have been a religious name deriving from the cluster of associations in the word and the suggestions of the flower: God made flesh, the Passion, drops of Christ's blood, the colour and sweet scent; all of which would have commended itself less after the Reformation, when *Incarnacyon*, as it seems, was shortened to *Carnation* as if in reference only to the flesh-

7. Carnations (Besler, *Hortus Eystettensis*, 1613)

colour of the flowers. (Turner 1562 also writes of 'incarnation roses' – i.e. flesh-coloured roses.)

CARRAGEEN (the common edible seaweed *Chondrus crispus*). XIX century. Shortened from Carragheen Moss, usually explained as seaweed associated with its presence at Carragheen, Co. Waterford, Ireland; but no such place exists. Probably Anglo-Irish *carrigeen*, from Irish *carraigín*, 'little rock (plant)', since *C. crispus* grows on rock.

CARROT (*Daucus carota* var. *sativa*). XVI century. Via the Old French *carotte* from Latin *carota* (as in the cookery book *De Re Coquinaria*, ascribed to Apicius), from Greek *karoton*.

CASHEW (*Anacardium occidentale*, native of tropical America; and its nuts). XVIII century. Called by the Tupi of eastern Brazil *acaju*, whence Portuguese *cajú*, English *cashew*. The nuts were known to 16th-century botanists as *Elephanten laüse*, 'elephant-lice'.

CASSAVA (*Manihot esculenta*, native of Brazil). XVII century. From the Portuguese *cassave*, from *casavi* in the language of the Taino people of Haiti.

CASTOR OIL PLANT (*Ricinus communis*, Palma Christi, native of Africa, introduced in the 16th century). The seeds give 'castor oil'; but the name derives from Spanish *agno casto* by which the plant was known in the West Indies, as if a kind of Agnus Castus (*Vitex agnus-castus*), 'Casto oil' having been changed euphoniously to 'Castor oil'. See also *Agnus Castus*.

CASUARINA (trees of the East Indian and Australian genus *Casuarina*). XIX century. English use of the botanists' Latin, from the Malaysian name for the Casuarina, *pohon kasuari*, supposed in turn to have been named from the resemblance of its peculiar branches to the feathers of the cassowary, Malay *kesuari*.

CATALPA (trees of the American genus *Catalpa*, especially *C. bignonioides*, the Indian Bean, introduced in the 18th century).

XVIII century. From the name of *C. bignonioides* in the language of the Creek Indians, *katuhlpa*, 'head without wings'.

CATAPUCE (in Chaucer). See *Spurge*.

CATCHFLY (*Silene armeria* of gardens, and other sticky species of *Silene*, and *Viscaria*). Name given by Gerard 1597 to *S. armeria*. See *Nottingham Catchfly*.

CAT GUT (the seaweed *Chorda filum*, Sea Laces, Dead Men's Ropes). XIX century.

CATKIN (of hazel, willow, poplar, etc.). For catkins there was no English word until Lyte 1578 invented catkin for the Dutch *katteken*, literally 'kitten'.

CATMINT (*Nepeta cataria*, native of Southern Europe and Western and Central Asia, introduced in the Middle Ages). XIII century. *Mint*, for the pungency of its scent. Anciently associated with cats who show a liking for it, and known by such medieval Latin names as *herba felina*, *herba cati*, (*herba*) *cattaria*. Albertus Magnus in the 13th century wrote that cats were supposed to become pregnant by means of catmint, a statement repeated in the 15th century herbal, Stockholm MS. X 90: 'the vertu of this herbe is gef a cat ete ther-of it schal conseywyn and brynge forth kytlyngis anon.'

CATNEP, CATNIP (*Nepeta cataria* – see *Catmint*, above). *cat* + *nepte*, Old English *nepte*, from medieval Latin *nepta*, from Latin *nepeta*, a mint-like plant.

CAT'S-FOOT (*Antennaria dioica*). XVIII century. From the soft woolliness of the leaves, translating the German *Katzenpfötchen*, 'cat's little paws'.

CATTLEYA (orchids of the genus *Cattleya*, from tropical America, of which the first was introduced in 1815). 1846. English use of the botanical Latin *Cattleya* (1824) given in honour of the horticultural patron William Cattley (d. 1832). (Cf. Proust's *Du Côte de Chez Swann* 1913, trans. by C. K. Scott Moncrieff 1922, *Swann's Way*, Part Two, Odette wearing cattleyas, and the phrase 'Do a cattleya' employed by Swann and Odette: 'The metaphor "Do a cattleya", transmuted into a simple verb which they would employ without a thought of its original meaning when they wished to refer to the act of physical possession.')

CAULIFLOWER (*Brassica oleracea* var. *botrytis*). XVI century, *Cole Florie*, *Coleflorie* (Gerard 1597), adapted from Italian *cavolfiori*, 'flowers of cole', or cabbage, called by the early botanists *brassica cauliflora*.

CAYENNE (red pepper from dried fruits of *Capsicum frutescens*, native of the Tropics). XVIII century. From the Tupi-Guarani *kyinha*, '(of) Cayenne island'.

CEANOTHUS (species and garden hybrids of the genus *Ceanothus*, mainly from North America). XIX century. From modern botanists' Latin, Linnaeus having borrowed his name for these North American shrubs from *keanotos* or *keanothos*, a plant mentioned by Theophrastus.

CEDAR (*Cedrus libani*, Cedar of Lebanon, native of Asia Minor, introduced 1683; and its timber). XIII century. From Old French *cedre*, from Latin *cedrus*, from Greek *kedros*, from an Indoeuropean *ked-*, to smoke. But the *cedrus* or *kedros* of the Ancients was juniper, as a rule the small Mediterranean tree or shrub *Juniperus oxycedrus*, which Calypso was very fragrantly burning in her cave, in the *Odyssey*, and which Virgil mentions for smoking out stables.

CELANDINE (Greater Celandine, *Chelidonium majus*; Lesser Celandine, *Ranunculus ficaria*). XIV century, *celidoine*. From Old French *celidoine*; from medieval Latin (*herba*) *chelidonia* or *celidonia*, or *chelidonium*; from Latin *chelidonia*; from Greek *khelidonion*, from *khelidon*, 'a swallow'. The ancient explanation of the name of the Great Celandine was twofold – that its flowers began and ceased with the coming and going of the swallow, and that it was swallow plant because swallows applied it to restore sight to their young, when they were blinded or eyeless. The use of *C. majus* for the eyes is certainly older than the explanation or the civilization of the Greeks: it was equally a part of ancient Chinese medicine. Pliny and Dioscorides distinguished a great and a small *chelidonia* or *chelidonion*: our *C. majus* and *R. ficaria*, or Lesser Celandine.

CELERIAC (*Apium graveolens* var. *rapaceum*). The root variety of Celery (q.v.).

CELERY (*Apium graveolens* var. *dulce*). XVII century. From French *céleri, céleri d'Italia*, from Italian *selleri*, plural of *sellero*, the word used on the plain of the Po; from Latin *selinum*, from Greek *selinon*. As an English word, *sellery*, first printed by John Evelyn in 1664, by which year the name and the vegetable were evidently familiar, though it took time for the English spelling to settle down; in the 18th century it was *celeri, salary, Italian celeri*.

Celeriac, XVIII century, seems an invention of English gardeners to distinguish the rooted variety, as if there had been a form *celeriacum*.

CENTAURY (*Centaurium minus*). XIV century. From medieval Latin *centaurea*, from Latin *centaureum, centaurion*, from Greek *kentaurion, kentaureion*, the 'centaurian (herb)'. *C. minus* was taken to be the 'gall of the earth' of the Romans (in fact a gentian?), which Pliny describes as a small leaved *centaureum*, in contrast to the larger *centaureum* (? the rather beautiful Yellow Knapweed, *Centaurea salonitana*, of the eastern Mediterranean), with which it was said that Chiron the centaur, one of mythology's divine healers or medicine men, was cured after allowing an arrow belonging to Hercules to fall on his foot.

CHAMOMILE (*Anthemis nobilis* and *Matricaria chamomilla*, wild Chamomile). XIV century. From French *chamomille*, from medieval Latin *chamomilla*, from Latin *chamaemelon*, from Greek *chamaimelon*, 'apple on the ground', on account of the apple scent of the flower-heads. *A. nobilis*, by origin a Western European species, was taken to be the *chamaimelon* of Dioscorides; which was probably the Wild Chamomile, in which the apple scent is more pronounced.

47

CHANTERELLE (the edible fungus *Cantherellus cibarius*). XVIII century. French *chanterelle*, from botanists' Latin *cantharellus*, a small *cantharus* or drinking-cup.

CHARD (*Beta vulgaris* var. *cicla*, White Beet, Beet Chard, Swiss Chard). XVII century. From the French *carde*, from Latin *cardus, carduus*, 'thistle'. First used in English for the edible part of an artichoke, then for the white edible rib of Chard. See *Cardoon*.

CHARLOCK (*Sinapis arvensis*). Old English *cerlic*, **cearloc*, one of several plant names with the suffix *-oc, -uc*. An old name as befits a weed of farm land, but of unknown significance.

CHASTE TREE (*Vitex agnus-castus*, Agnus Castus, from Southern Europe, introduced in the 16th century). XVI century, Turner 1562, translating the Latin *castus*. See *Agnus Castus, Castor Oil Plant*.

CHEDDAR PINK (*Dianthus gratianopolitanus*). XVIII century. *D. gratianopolitanus* was first discovered in England as a wild plant on the sides of the Cheddar Gorge in Somerset (its only English station), c. 1723, by the Wiltshire botanist Samuel Brewer (1670–1743).

CHERRY (cherry species of *Prunus*, and their derivatives; and fruit). XIV century. From the Old Norman French *cherise* (as if *cherise* was a plural). This and the Old English *ciris, cyrs*, go back to Latin *cerasus*, Greek *kerasos*. Pliny wrote that they had no cherry trees – i.e. cultivated kinds – until the war against Mithridates VI, King of Pontus, which began in 75 BC. Lucullus brought cherry trees back after the king's defeat – 'and within 120 years cherry trees had crossed the sea as far as Britain'. The Romans derived the name from the Pontine city of Cerasus, the Turkish Gireson, on the Black Sea. But this was folk-etymology, *kerasos* going back to the Accadian *karsu*.

CHERRY-LAUREL (*Prunus laurocerasus*, Laurel, native of S.E. Europe and S.W. Asia, introduced early in the 17th century). XVII century, Evelyn 1664. From its cherry-like fruits.

CHERVIL (*Anthriscus cerefolium*, native of S.E. Europe and Western Asia, introduced in the Middle Ages). Old English *cerfille*, from Latin *caerefolium*, from Greek *khairephullon*, 'leaf of gladness', or 'greeting' (*phullon*, 'leaf', *khairein*, 'to be glad', 'to welcome').

CHESTNUT (nut of *Castanea sativa*; and the tree, native of Southern Europe, North Africa, and S.W. Asia, introduced, but planted as early as Anglo-Saxon, or probably Roman times). In Old English a chestnut tree was *cist, cisten, cistel, cist-bēam*, words which occur in place-names. Chestnut is 14th century *chesteine*, from Old French *chastaigne*, + nut; both *cist*, etc., and *chesteine* going back to Latin *castanea*, from Greek *kastanea*, 'chestnut-tree' (chestnuts were *kastana*) – all as if from the city of Kastana in Northern Greece, or a city in Pontus on the Black Sea named *Kastanis*, derivations which are ancient folk-etymology for a loan-word, cognate with the Armenian *kaskeni*, 'chestnut-tree'. The cities were no doubt named from the tree or its nuts.

CHICKWEED (*Stellaria media*). xv century. Eaten by, and fed to chickens, goslings, cage-birds. In early botanists' Latin *morsus gallinae*, 'hen's bite'.

CHICORY (*Cichorium intybus*; also *C. endivia*, Endive, native of India or Egypt, introduced in the 16th century). xv century, *cicoree*, from French *cichorée*, from medieval Latin *cichorea*, from Latin *cichoreum*, from Greek *kikhorion*, from Egyptian *kehsher*. See also *Endive*.

CHILLI (dried fruits of *Capsicum frutescens*, native of the Tropics). xvii century. From Spanish *chile*, from *chilli* in Nahuatl, the language of the Aztecs.

CHINA ASTER (*Callistephus chinensis*, introduced from China in 1731). xviii century. It was known at first as *Aster chinense*. See *Aster*.

CHINA ROSE (*Rosa chinensis*). xviii century. Garden forms of *R. chinensis* were introduced from China in the second half of the 18th century.

CHINESE CABBAGE (*Brassica sinensis*, Pak-choi; and *B. pekinensis*, more usually known as Pe-tsai. Both introduced 1770). XIX century.

CHIVE, CHIVES (pl.) (*Allium schoenoprasum*). XIV century. From the Old French *cive*, from the Latin *cepa*, 'onion'.

CHOCOLATE (beverage from the burnt and powdered beans of the Cacao Tree, *Theobroma cacao*, native of Central and South America, first encountered by Europeans in Mexico). XVII century. From the French *chocolat*, from Spanish *chocolate*, from the Nahuatl *xocoatl*.

CHRISTMAS ROSE (*Helleborus niger*, native of Central and S.E. Europe, introduced at the close of the 16th century). XVII century. Also *Christmas Flower*; and *Christmas Herb*, XVI century. Cf. Culpeper 1653 (who calls it, inter alia, by the second and third of these names): 'about Christmas time, if the weather be anything temperate, the flowers appear'.

CHRIST'S THORN (the thorny shrub *Paliurus spina-christi*, from Southern Europe to China, introduced late in the 16th century). XVI century, Turner 1562, translating the botanists' Latin *spina Christi*. From its abundance in Palestine, *P. spina-christi* was supposed to have furnished Christ's crown of thorns.

CHRYSANTHEMUM (for *Chrysanthemum morifolium*, introduced from China in 1764, and its forms). XVIII century. Chrysanthemum, from the Greek *khrusanthemon*, 'gold' + 'flower', was used in botanists' Latin and English, in the 16th century, for the Corn Marigold (*C. segetum*). Linnaeus adopted *Chrysanthemum* as the generic name, which was taken over in speech for the rapidly popular *C. morifolium* from China.

CINERARIA (the Cineraria of the florists, derived from species of *Senecio* introduced from the Canaries in the last decades of the 18th century). English use of the botanists' Latin, from *cinerarius*, 'ashy', and originally the 16th century name for Dusty Miller, *S. cineraria* of gardens, on account of its silvery-white down.

CIDER (fermented apple juice). XIV century. From Old French *sidre* with the same specialized meaning, from Late Latin *sicera*, from the Biblical Greek *sikera*, reproducing the Hebrew *shēkhār*, intoxicating drink, from *shākhar*, 'he became drunk', cf. Accadian *shakaru*, to be drunk.

CINNAMON (inner bark of the Cinnamon Tree, *Cinnamomum zeylanicum*, native of India and Malaysia). XV century, *sinamome*. From French *cinnamome*, from Latin *cinnamomum*, from Greek *kinnamumon*, *kinnamon*, from the Hebrew *qinnāmōn*. Fabulous accounts of cinnamon are given by Herodotus – 'the dry sticks the Phoenicians have taught us to call cinnamon' – and Pliny.

CINQUEFOIL (*Potentilla reptans*). XVI century (earlier 'fyfleved gras'; Old English *fifleaf*). Taken to be the Dioscoridian herb of ancient apotropaic and medicinal power, *pentadaktulon* ('five finger'), *pentaphullon* ('five leaf'); in Latin the *quinquefolium*, from which Cinquefoil derives, via French. The leaves of *P. reptans* have mostly five leaflets; and the flower has its parts in five.

CITRUS (collectively, for species and fruits of *Citrus*). XIX century. English use of the Linnaean generic name, from the Latin *citrus*, the citron tree, *C. medica*, from the Far East (which is the source of candied peel).

CLARET (red wines from the Bordeaux region). XIV century. From Old French (*vin*) *claret*, 'clear wine', 'clarified wine'.

CLARKIA (species of the genus *Clarkia*, native in Pacific North America. *C. pulchella* was introduced in 1826). English use of the botanists' Latin, 1814, given in honour of Captain William Clark (1770–1838), joint leader of the Clark and Lewis expedition which crossed the Rockies to the Pacific, 1803–1806.

CLARY (*Salvia sclarea*, native of Southern Europe). XIV century. Old English *slarie*; 15th century *clary*. From the medieval Latin name *sclarea*, which is unexplained.

CLEAVERS (*Galium aparine*). XV century *clivre*, that which cleaves or sticks when it comes up against you. Cf. Dutch *kleef*,

klevers, and related names in German. The Old English name for
G. aparine was *clīfe* with the same meaning (*clifian*, 'to stick'), used
also for the Burdock (*Arctium lappa*).

CLEMATIS (for the garden species, and forms, of the genus
Clematis). XIX century. English use of the Linnaean name of the
genus. Clematis, from Latin *clematis*, Greek *klematis*, from
klema, 'vine-shoot', was the botanists' Latin for *C. vitalba*, Old
Man's Beard, and in that sense was already in use as an English
word in the 16th century.

CLEMENTINE (citrus fruit of a hybrid, raised in Algeria c.
1900, between ? the Mandarin, *Citrus nobilis* var. *deliciosa*, and
the Seville Orange, *Citrus aurantium*). From French *clémentine*,
named after the priest Father Clément who raised the hybrid
near Oran.

CLOTE (*Nuphar lutea*, Yellow Water-lily, Brandy-bottle). Old
English *clāte*, originally for the Burdock, *Arctium lappa*, the basic
meaning of which is something which sticks, i.e. which has burs.
The name was extended to other plants with broad leaves,
including the Yellow Water-lily, Coltsfoot (*Tussilago farfara*) and
Butterbur (*Petasites hybridus*).

CLOUDBERRY (*Rubus chamaemorus*, plant and fruit). XVI
century. Gerard 1597 gave a folk-etymological explanation, from
the presence of *R. chamaemorus* on top of Ingleborough and
Pendle, 'two of the highest mountaines in all England, where the
clouds are lower than the tops of the same all winter long, whereupon the people of the countrie have called them Cloud berries.'
In fact, from Old English *clūd*, 'rocky hill', + berry. (Cf. 16th
century *knotberry* for *R. chamaemorus*, from English *knot*, 'rocky
hill', from Old Norse *knott*.)

CLOVE-PINK (*Dianthus caryophyllus*, native of Southern Europe,
introduced in the Middle Ages). XIX century. See *Cloves, Gilly-
flower, Pink*.

CLOVER (species of the genus *Trifolium*). Old English *clæfre*,
with cognate names in other Germanic languages.

CLOVES (spicy dried flower-buds of *Eugenia aromatica*, the Clove-Tree of the Moluccas). XIII century. From Old French *clou*, 'nail', the flower-buds having the shape of small medieval nails. Originally (an involved story with its peculiar sequel: see *Gilly-flower*) *clowe gilofre*, from French *clou de girofle* – *girofle* from Pliny's *caryophyllum*: 'Also in India they have a grain like pepper, though larger and more fragile, which they call caryophyllum . . . It is traded for its scent.' Pliny's word *caryophyllum*, as if Greek 'nut-leaf', appears to be an ingenious and appropriate folk-etymological interpretation of an Indian name of the Clove-Tree, *katakaphalam* = 'whose fruit is pungent'. The double English word was shortened, as we have it, by the 14th century.

CLUBMOSS (species of the genus *Lycopodium*). XVI century. Translation of apothecaries' Latin *muscus clavatus*.

COB-NUT (*Corylus avellana* var. *grandis*, cultivated Hazel, and nut). XVI century. *cob* + *nut*. Cf. 15th century 'cobill nut', as if like a cobble or rounded stone, larger than the wild nut. But from the 16th century at least the children's game 'cobnut' was played with hazel-nuts. In this predecessor of Conkers (see references in the *Oxford English Dictionary* and Iona and Peter Opie's *Children's Games in Street and Playground*, 1969) the winning nut was the 'cob'. See *Conker*.

COCKSCOMB (the seaweed *Plocamium coccineum*). XIX century. From its rose-red colour of a cock's comb.

COCKSCOMB (*Rhinanthus minor*, Yellow Rattle). XVI century, Lyte 1578, translating the Dutch name: 'of some Hanekamme-kens, that is to say, Hennes Commes, or Coxecombes'; from its resemblance, despite the yellow flowers, to the Red Rattle (*Pedicularis palustris*) with its cock's-comb-coloured flowers.

COCKSCOMB (*Celosia argentea* var. *cristata*, of tropical origin, introduced in the 16th century). XVIII century. From the dark-red chaffy spikes, reared like a cock's comb.

COCKSFOOT (the grass *Dactylis glomerata*). XVII century. Describing (not very well) the branched panicle.

COCOA (beverage from powdered seeds of the tree *Theobroma cacao*, the Cacao Tree, native of Central and South America, first encountered by Europeans in Mexico). XVI century, *cacao*, for the seeds or beans; early XVIII century *cocoa*, for the same; late XVIII century for the powder and the beverage. From Spanish *cacao*, from the Nahuatl *cacahuatl*, 'cacao beans'.

COCONUT (nuts of the species of trees formerly grouped in the genus *Cocos*; and the tree). XVII century. In English the nut (which the Portuguese were bringing back in quantity from the Maldive Islands) was first called *cocus, coco* (16th century). From Portuguese *coquos*, late 15th century, from *côco*, 'grinning face', 'bogeyman' (cf. Coco the Clown), in reference to the face-like appearance of the nut with its shaggy hair and round depressions like eyes and mouth. The plural *coquos* was Latinized into *Cocus*.

CODLIN, CODLING (kinds of hard cooking apple). XV century, *querdling*. From Anglo-Norman *quer-de-lion*, from French *coeur-de-lion*, 'lion's heart'.

CODLINS-AND-CREAM (*Epilobium hirsutum*). XVII century. Recorded by John Ray 1670 and explained by him as 'from the smell of the leaves a little bruised'; but the bruised leaves have no suggestion of apples (codlins), or stewed apples with cream. The name perhaps derives from a corruption of Gerard's 'Codded Willow herbe' (i.e. with flowers on a codd or stem), as if 'codlin willow herb'; this in turn having suggested 'codlins-and-cream' from the rosy petals and creamy-white stigmas. A further suggestion of coddling (=stewing) may explain other vernacular names for *E. hirsutum*, which have to do with various fruit, apples included.

COFFEE (the drink, etc., from roast beans of species of the tropical African genus *Coffea*). XVII century. First popular in England in the sixteen-fifties and sixties, the drink was mentioned as 'chaova' in 1598, 'coffe' in 1601; later 'cahve', 'cauphe', etc. From Turkish *qahve*, from Arabic *qahwah*, probably from the district of Kaffa, in the Ethiopian highlands, one of the original sources, where the wild coffee-trees are still exploited.

COLE (various kinds of cabbage). Old English *cāl*, contracted form of *cawel*, from the Latin *caulis* (Greek *kaulos*), 'stem', 'stalk' (of a cabbage). The basic meaning is 'that which is hollow'.

COLOCYNTH (*Citrullus colocynthis*, native of the Mediterranean region). XVII century. From Latin *colocynthis*, from Greek *kolokunthis*.

COLTSFOOT (*Tussilago farfara*). XVI century. From the leaf shape, translation of the medieval Latin name *pes pulli*, 'foal's foot'. *T. farfara* was the *pes pulli agrestis*, 'foal's foot of the land', in contrast to the Yellow Water-lily, *Nuphar lutea*, the *pes pulli aquaticus*, 'foal's foot of the water'.

COLUMBINE (*Aquilegia vulgaris*). XIII century. From the Old French *colombine*, from medieval Latin *columbina* (*herba*), 'dove-like herb', from Latin *columba*, 'dove', describing the extraordinary likeness of the spurred petals of each flower to a quintet of doves or pigeons.

COMFREY (*Symphytum officinale*). XV century. Comfrey was one of the chief plants in early medicine for wounds and fractures, the *consolida major*, identified with a wound-herb which Pliny mentions as growing in running water particularly in the Alps – the *conferva* (*confervere*, 'to grow together', 'consolidate'). So we have Comfrey (*Conforye*, 15th century) by way of medieval Latin and French.

COMICE (variety of pear). XX century. Shop name shortened from *Doyenné du Comice*, q.v.

CONKER (shelled seeds of the Horse-chestnut). XX century in this sense, apparently short for 'conqueror', from the children's game 'Conquerors', in which the warring object was formerly a cob-nut (q.v.). The *Oxford English Dictionary* identifies conker as a dialectical word for a snail-shell, a conquering game (of a different kind) having been played with snail-shells which the children called conkers. But this is unconvincing (see I. and P. Opie, *Children's Games in Street and Playground*, 1969, in which 1848 is given for the earliest mention of playing 'conquerors' with horse-chestnuts).

COPPER-BEECH (*Fagus sylvatica* var. *atropunicea*). XIX century. Describing the burnished copper colour of its leaves (especially the young leaves).

COPRA (dried kernels of the Coconut, q.v.). XVI century. From the Portuguese *copra*, from the Malayalam *koppara*, 'coconut'.

CORAL-ROOT (*Corallorhiza trifida*). XIX century. Translation of the 18th century generic name describing the creamy-coloured branched rhizome, likened to coral.

CORIANDER (*Coriandrum sativum*, native of Southern Europe). XIV century. From the Old French *coriandre*, from Latin *coriandrum*, from Greek *koriannon* (also *korion*), perhaps from Greek *koris*, 'bed-bug', on account of the scent and appearance of coriander seeds.

CORK (bark of the Cork Oak, *Quercus suber*, native of the Mediterranean region). XIV century. From the Spanish *alcorque*, 'cork-soled slipper', from Spanish-Arabic *alqurq* (*al*, the definite article), from Latin *quercus*, 'oak'.

CORN (collectively for kinds, or fields, of cereal). Old English *corn*; which was already used of the seed, not the standing crops. With the related Germanic words, 'corn' goes back to an Indo-european base meaning to wear down, wear away; so 'corn' was used as we now use 'grain' (which is a cognate word): e.g. a 'corn' of salt, or sand; cereals, no less than pepper and other plants, producing corns or grains. The extra-specialized use of corn in American English for Indian Corn, *Zea mays*, began in the 18th century.

CORN COCKLE (*Agrostemma githago*). XVIII century. Earlier and generally *Cockle*, Old English *coccel*, for which the *Oxford English*

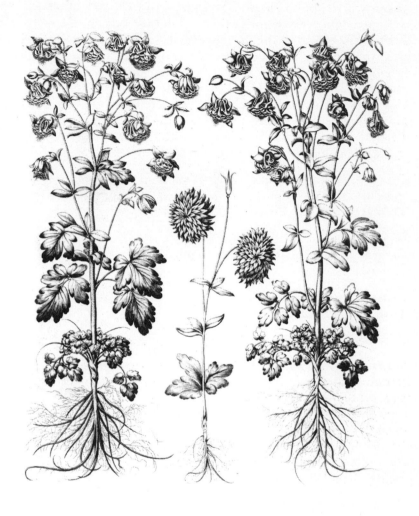

8. Columbines (Besler, *Hortus Eystettensis*, 1613)

Dictionary hazards a derivation from Latin *coccus, coccum,* the colour scarlet; but perhaps to be connected with cock, the bird, as from a diminutive of medieval *coccus* in that sense, in reference both to the colour of the flowers and the cock's comb, and the way the flowers top the corn. Cf. French *cocquelicot,* which originally meant cock, cockerel, and German *Gockelar,* for poppy; for which English local names include *Cockeno, Cock's Comb, Cock's Head.* For Corn Cockle there are names in German and Slavonic languages connecting it with the bird, German *Gugol, Kuckel* (=*Gockel,* cockerel, chanticleer), Czech *koukol.* And cf. in East Anglian *Cockerel* for Corn Cockle.

CORNFLOWER (*Centaurea cyanus*). XVI century, Lyte 1578, as translation of the botanists' or apothecaries' Latin *flos frumenti.* The earlier names were *Bluebottle,* or *Blue Bothem* (15th century).

CORN MARIGOLD (*Chrysanthemum segetum*). XVI century, Gerard 1597. The commoner name was Gold (q.v.). See *Marigold.*

CORN SALAD (*Valerianella locusta*). XVI century, Gerard 1597. Then, as now, eaten as a salad.

CORN SPURREY. See *Spurrey.*

CORNEL (*Cornus mas,* the Cornelian Cherry of Southern Europe, introduced to English gardens at the end of the 16th century). XVI century. From the German *Kornel(baum),* cornel-(tree), from medieval Latin *cornolium,* the tree producing *cornolia,* i.e. the Latin *corna,* the brilliant scarlet drupes of the *cornus,* the Cornelian Cherry Tree, from Greek *kranon* or *kraneia.* (When Circe changed the followers of Odysseus into pigs, in the *Odyssey,* she fed them on acorns, beech-mast, and cornelian cherries.)

CORNISH MONEYWORT (*Sibthorpia europaea*). XIX century. Cornish, because first recorded as an English plant in Cornwall, by John Ray, between St Ives and Land's End, 1 July 1662; Moneywort (q.v.), from having more or less coin-shaped leaves resembling those of the Moneywort, *Lysimachia nummularia.*

58

Cos Lettuce (*Lactuca sativa* var. *longifolia*). XVII century (John Evelyn, in his *Acetaria. A discourse of sallets*, 1699). Variety of lettuce from the island of Cos. See *Lettuce*.

Costmary (*Chrysanthemum balsamita*, Alecost, of Western Asian origin, introduced by the 16th century). XV century. *cost + Mary* the Cost of the Virgin Mary. Cf. the medieval Latin name *Herba Sanctae Mariae*. See *Alecost*.

Cotoneaster (species of the genus *Cotoneaster*, many of which were introduced from Nepal into English gardens early in the 19th century). XIX century. English use of the Latin generic name, from the 16th century botanists' Latin *cotonaster*, from Latin *cotoneum*, 'quince', + suffix *-aster*, 'rather like a quince-tree'.

Cotton (seed down from species of the genus *Gossypium*, natives of Asia). XIV century. From the Old French *coton*, from the Genovese Italian *cottone*, from the Arabic *qutun*. In several European languages cotton is known by an equivalent to 'tree-wool'. Pliny writes of the cotton plant, Chinese, Arabian and Egyptian, as (*arbor*) *lanigera*, 'wool-bearing tree'; also as *gossypinus*, from Bahrein.

Cottonwood (*Populus deltoides*, native of North America, introduced in the 18th century). XIX century. From the cottony seeds.

Couch, Couch-grass, Quitch (*Agropyron repens*). Old English *cwice*, from *cwicu*, 'living', 'lively'. A plague still of farming and gardening, *A. repens* has cognate names in other Germanic languages, e.g. German *Quecke*; and *Quitz* (Schleswig-Holstein).

Couscous, Cuscus (the cereal *Pennisetum glaucum*, grown in North Africa; and its grains. Pearl Millet, African Millet). XVII century. From Arabic *kuskus*, 'that which is pounded' (*kaskasa*, 'he pounded').

59

COWAGE (hooked hairs, which cause itching and were used in honey against intestinal worms, from the pods of the tropical climber *Mucuna pruritum*; and the plant). XVII century, Parkinson 1640. From the Hindi *kavac*.

COWBANE (*Cicuta virosa*). XVIII century. A plant which poisons grazing cattle.

COWBERRY (*Vaccinium vitis-idaea*; and fruit). XIX century. English equivalent for the generic Latin name *vaccinium*, 'pertaining to cows'.

COW-PARSLEY (*Anthriscus sylvestris*; also, without popular distinction, *Chaerophyllum tremulum*, and *Torilis japonica*). XVIII century. I.e. plant like an inferior parsley, fit only for cows, or found along lanes and cow-paths.

COW-PARSNIP (*Heracleum sphondylium*, Hogweed). XVI century. Coined by Turner 1548. 'Sphondilium . . . It may be called in englishe Cowpersnepe.'

COWSLIP (*Primula veris*). Old English *cūslyppe*, *cūsloppe*, 'cow slime', 'cow-dung', 'cow-pat', as if in the sense of growing wherever a cow-pat has fallen. *Cowslop* still in some dialects. The politer and seemingly inoffensive *cowslip* became standard in the 16th century (e.g. with Shakespeare), as *slip* in the sense of cow dung became obsolete.

COW-WHEAT (species of the genus *Melampyrum*). XVI century, Lyte 1578, translating the German peasants' name *Kuweyssen*, for the showy and beautiful (and in England rare) *M. arvense* which grows on the balks and edges of corn fields; 'cow-wheat' because freely cropped by passing cattle, to which it was also fed in times of scarcity.

COX'S ORANGE PIPPIN. Apple variety raised in 1830 by Richard Cox (c. 1776–1845), at Colnbrook near Slough, and first popularized by an award at the Grand Fruit Exhibition of the Horticultural Society in 1857. See also *Pippin*.

9. Cowslip (Fuchs, *De Historia Stirpium*, 1542)

CRAB (-APPLE; -TREE) (*Malus sylvestris*). XV century. Of un-known but presumably Scandinavian origin. In the north of England *Scrab* (cf. Turner, *Herbal* 1562: 'In the Southe countre a Crab tre, in the North countre a Scrab tre').

CRACK-WILLOW (*Salix fragilis*). XVII century, Ray 1670. Describing the way the branches snap or crack off.

CRANBERRY (*Oxycoccus palustris*; and fruit). XVII century; apparently popularized with the importation of cranberries from New England (fruit of *O. macrocarpus*); but there is no reason to suppose it was not an English name, if unrecorded, for a once much commoner marshland plant and fruit – growing where cranes (also once common in England), or cranes = herons, were to be found. There are similar names in German.

CRANESBILL (various species of the genus *Geranium*). XVI century. Turner 1548, translating the German *Kranchschnabel*, which describes the long beak of the carpel. (Geraniums were taken to be the *geranion* – Greek *geranos*, 'crane' – of Dioscorides and Pliny.)

CREEPING JENNY (*Lysimachia nummularia*). XIX century.

CRESS (especially Watercress, *Nasturtium officinale*; and Garden-cress, *Lepidium sativum*, one of the more ancient salad plants, a native probably of Western Asia). Old English *cærse, cerse, cresse*, with cognate names in other Germanic languages, from an Indo-european base meaning to nibble or eat. As *N. officinale*, often in place-names, such as Cresswell, Kerswell, 'cress brook', 'cress spring'. In Old English *N. officinale* was distinguished as *ea-cerse*, 'stream cress', and *wielle-cerse*, 'spring' or 'brook cress'.

CROCUS (species of the genus *Crocus*). XVII century. English use of the Latin *crocus*; from Greek *krokos*, the Saffron Crocus (*C. sativus*, native of Asia Minor), which was a Semitic loan-word.

CROSSWORT (*Galium cruciata*). XVI century, Lyte 1578, translat-ing the botanists' Latin *cruciata* (*planta*), 'crossed plant', i.e. with leaves in a cross-like arrangement.

CROWBERRY (*Empetrum nigrum*). XVI century, Gerard 1597. There are parallel names in German, and in Scandinavian languages. Objurgatory: the black fruit is a meal for black crows rather than humans.

CROWFOOT (the 'Buttercup' species of the genus *Ranunculus*). XV century, when several kinds are mentioned in manuscript herbals, though not clearly distinguished. Since crowfoot has remained a book name until the present day, it was perhaps no more than a translation of the medieval Latin of the apothecaries, *pes corvi*. The crow's foot, not very strikingly, is the shape of the divided leaves. Cf. Latin *coronopus*, Greek *koronopous*, 'crow foot', taken usually to have been Buck's-horn Plantain (*Plantago coronopus*).

CROW-LEEK, CROW-GARLIC (*Allium vineale*). Old English *crāwan lēac*; XIV century *crawegarleke*. A leek or garlic for crows. Objurgatory, and with reason, as a farm weed able to give a garlic flavour to milk and butter. See *Garlic, Leek*.

CROWN IMPERIAL (*Fritillaria imperialis*, from Persia, introduced at the close of the 16th century). XVI century. In reference to the crown-like and more than royal appearance of the hanging flowers and tuft of leaves above them, translating the Latin *corona imperialis*, the name 16th century botanists gave to *F. imperialis*, which was sent from Constantinople to Vienna in 1575. The plant became rapidly popular with gardeners, and the name with poets, George Chapman, 1575 ('Fayre Crowne-imperiall, Emperor of flowers'), Shakespeare, 1611 (as one of the flowers let fall by Proserpina from Dis's waggon, in the speech by Perdita, in *The Winter's Tale*), Ben Jonson, George Herbert, etc. Shakespeare's Crown Imperial suggests that he knew the name better than the flower. Crown Imperials are a bit large and formal even for a goddess to go picking with daffodils, violets, primroses, oxlips, etc.

CUCKOO-FLOWER (*Cardamine pratensis*). XVI century. The commonest name for *C. pratensis*, and though it was no doubt in use before the first record of it (Lyte 1578) it probably came

63

into plant books as a translation of the German *gauchbloom*, recorded in 1539, or of the herbalists' Latin *flos cuculi*. 16th century explanations of Cuckoo-flower are that *C. pratensis* blossoms with the arrival of the cuckoo, and that it is frequently enveloped in the foam (around frog-hopper nymphs) which was held to be ejaculated by cuckoos – 'cuckoo-spit'.

CUCKOO-PINT (*Arum maculatum*). XV century, *cokkowyl pyntyl, cokkupyntel*; XVI century, Turner 1551, in the shortened form *coccowpynt*. *Pintle*, i.e. cock, prick, from the erect spadix. The sense is the erect pintle of the lustful cuckoo (this being a flower blossoming with the arrival of the cuckoo) coming to father birds who ought not to be in the nest. Also *Priest's-pintle, Priest's-pilly*, for the parson as the traditional village cuckoo (so in other languages, the pintle of priest or monk – cf. German *Pfaffenpint*, 'parson's pintle', in 16th century herbals; which may explain Turner's polite curtailing of *pintle* to *pint*).

CUCUMBER (*Cucumis sativus*, and fruit, of South Asiatic origin, and early cultivation). XIV century, *cocumer*, which in the 15th century was assimilated to the Old French *cocombre*. From the *cucumis, cucumeris*, of the Romans. Pliny has much to say of *cucumis*, which he describes as a favourite food of the Emperor Tiberius, whose gardeners raised cucumbers in glazed frames on wheels.

CUDWEED (species of the genera *Filago* and *Gnaphalium*). XVI century, Turner 1548, who gives it as a Yorkshire name for *G. sylvaticum*, distinguishing cudwort as *F. germanica*. Leonard Mascall in his *Government of Oxen* 1587 says that cudwort was given to cattle which had lost their cud.

CUMIN (*Cuminum cyminum*, native of the Mediterranean region, the seeds of which were imported in the Middle Ages). XII century. From the old French *comin*, from the Latin *cuminum*, from the Greek *kuminon*, a Semitic loan-word related to the Accadian *kamūnu*, literally 'mouse plant'.

CUMQUAT. See *Kumquat*.

CURRANTS *1*. (dried grapes). XIV century, *raysons of coraunce*, Old French *raisins de Corinthe*, raisins from Corinth, in Greece. By the early 16th century shortened to the final word. *2*. (fruits of *Ribes rubrum* and *R. nigrum* (which were not anciently cultivated); and the plants. XVI century. Lyte 1578 writes of Red Currants as 'Bastard Corinthes'.

CUSTARD APPLE (fruit of *Annona reticulata*, native of tropical America). XVII century. Because of the custard-like pulp.

CUVIE (the seaweed *Laminaria hyperborea*). XIX century (a name from the Orkneys).

CYCLAMEN (species of the genus *Cyclamen*, from Europe and the Mediterranean region). XVI century. From the botanists' Latin *cyclamen*, from the classical Latin *cyclaminos*, from Greek *kuklaminos*, from *kuklos*, 'a circle', 'a round' (i.e. the round corm).

CYPHEL (*Cherleria sedoides*). XVIII century. From the botanists' Latin *cyphella*, from Greek *kuphella*, 'hollows of the ear', a name given to the Houseleek (*Sempervivum tectorum*) for its cup-like rosettes and transferred to this little mountain plant which grows in close cushions (perhaps with a knowledge that the Greek *kuphella* also meant 'clouds of mist').

CYPRESS (*Cupressus sempervivens*, native of Western Asia, introduced in the 16th century). XIII century. From the Old French *cipres*, from Low Latin *cypressus*, from Latin *cupressus*, from Greek *kuparissos*, a word like the tree, of exotic origin, possibly from the Hebrew *gōpher*. See *Gopherwood*.

D

DABBERLOCKS (the edible seaweed *Alaria esculenta*). XIX century. ? *dabby*, 'wet and clinging', 'flabby', + *locks*, tresses.

DAFFODIL (*Narcissus pseudonarcissus*). XVI century, before 1548, a form (the causation of which is not satisfactorily explained) of *Affodill*, owing to the supposition that *N. pseudonarcissus* was an asphodel (q.v.).

DAHLIA (species, and their forms, of the genus *Dahlia*, of which *D. pinnata* was introduced from Mexico in 1789). XIX century. English use of the Latinized generic name (1791) commemorating the Swedish botanist Andreas Dahl (1751–1789).

DAISY (*Bellis perennis*). Old English *dæges ēage*, 'day's eye'. XV century, *dayeseye*. Descriptive of the way the flower opens and folds morning and evening, and of its likeness to a small sun. Cf. the medieval Latin name for the daisy, *solis oculus*, 'sun's eye'.

DAMASK ROSE (pink and white, spicey-scented roses, reputedly introduced by the humanist and royal physician Thomas Linacre [1460?–1524] who was in Italy from about 1485 to 1491). Early XVI century. Translation of the apothecaries' Latin *Rosa damascena*, 'rose from Damascus'.

DAME'S-VIOLET (*Hesperis matronalis*). XVI century, Lyte 1578. Translated from the botanists' Latin *viola matrinalis*, which in turn translated a Greek name in Dioscorides for some such plant. Colour and scent justified 'violet'. 16th century botanists weakly explained *matronalis* by saying that women liked to have *H. matronalis* in their gardens. The Dioscoridean name indicates a plant used against the diseases of women.

DAMSON (cultivated form of *Prunus domestica* ssp. *insititia*; and fruit). XIV century, *damascene*. This plum was taken to be the

66

10. Daffodil, Fritillary, and Narcissus (Besler, *Hortus Eystettensis*, 1613)

pruna damascena of the Romans, 'named from Damascus in Syria' (Pliny).

DANDELION (*Taraxacum officinale*). XV century, *dendelyoun, dawndelyoun*, from French *dent de lion*, lion's tooth, translating the medieval apothecaries' name *dens leonis*, the lion's tooth being either the tap root, the florets, or the jagged edge of the leaves.

DANEWORT (*Sambucus ebulus*, Dwarf Elder). XVI century, Turner 1538 *danwort*; 1568 *daynwort*): 'stink plant', an apt name. Parkinson 1640 adds that *S. ebulus*, as a purgative of strength, produced the 'danes', or diarrhoea. An ingenious combination of historical fancy and folk-etymology produced the early 17th century explanation that *S. ebulus* was a plant which grew from the blood of Danes killed in the fighting of Dane against Anglo-Saxon. See Grigson, *The Englishman's Flora*, 1955; and 'dain', in the *Oxford English Dictionary*.

DAPHNE (garden species of the genus *Daphne*). XIX century. From the name given by Linnaeus, 1737 (Greek *daphne*, the laurel). See *Mezereon*.

DARNEL (*Lolium temulentum*). XIV century. A sufficiency of Darnel seeds ground in wheat-flour and then baked in bread, etc., produces drunken-like states of confused perception. So the plant (now rare in crops of corn) has been known in many languages by a name indicating this, such as German *Schwindel*, dizziness, French *ivraie*. Darnel (in Walloon *darnelle*) is a name seemingly of the Flanders area with a similar meaning, related to names in Danish and Swedish.

DATE (fruit of the Date-palm, *Phoenix dactylifera*, native probably of Western Asia and North Africa; and the tree). XIII century (dates were a medieval import). From the Old French *date*, Latin *dactylus*, Greek *daktulos*, 'a date'; the Greek – 'finger' or 'toe' originally – having come to mean, in this matter of the date or the dried date, a fruit shaped like a toe.

DEADLY NIGHTSHADE (*Atropa belladonna*). XVI century, Lyte 1578, though he uses *deadly* adjectivally rather than as part of a name, 'this deadly Nightshade'. Translation of the apothecaries' Latin *solanum mortale*, or *solatrum mortale*. See *Nightshade*.

DEAD MEN'S FINGERS (*Orchis mascula*, Early Purple Orchid). XVII century. The *locus classicus* for this name is *Hamlet* (IV, vii, 170–2), where Shakespeare includes Dead Men's Fingers among the flowers of Ophelia's suicide. It is the name used by 'cold maids': the 'grosser name' given by 'liberal shepherds' (and warm maids) would have been one of those for *O. mascula* as an aphrodisiac plant (which has the shape of an erection above the two root-tubers) such as *Stander, Standergrass, Cullions, Cocks, Pintel, Ram's Horn*. In *The Faithful Shepherdess*, 1629, by John Fletcher, Clorin excludes 'foul standergrass' from the herbs she has been gathering by moonlight – since it induces 'appetite'.

DEAD MEN'S ROPES (the seaweed *Chorda filum*, Sea Lace, Mermaid's Tresses). XIX century.

DEAD-NETTLE (*Lamium album*, etc.). XIV century. Self-explanatory: 'dead', without the ordinary life, or stinging quality, of the true nettle.

DEATH CAP (the mushroom *Amanita phalloides*). XX century. Mycologist's name devised to warn amateur mushroom collectors against an exceedingly poisonous species which often causes death. Cf. *Destroying Angel*.

DELPHINIUM (garden kinds of the genus *Delphineum*). XIX century. English use of the botanists' Latin, which was derived from the identification of larkspurs with the Greek *delphinion*, 'dolphin-like (flower)', the spur representing the long 'beaked' head of the dolphin.

DEPTFORD PINK (*Dianthus armeria*). XVI century, Gerard 1597. Gerard described a pink 'which groweth in our pastures neere about London, and other places, but especially in the great field next to Detford, by the path as you go from Redriffe to Greenewiche'. The details he gives prove that he was in fact describing the Maiden Pink, *D. deltoides*.

DERRIS-POWDER (powdered root, a contact insecticide, of species of the South Asiatic genus *Deguelia*). XX century. The genus was formerly known as *Derris*, botanists' Latin from the Greek *derris*, skin.

DESTROYING ANGEL (the mushroom *Amanita virosa*). XX century. *A. virosa* is entirely white, and extremely poisonous, as if Azrael, the Angel of Death (though Azrael only presides over death). A mycologist's invention, as a warning (cf. Death Cap).

DEVIL'S-BIT (*Succisa pratensis*, Devil's-bit Scabious). XV century. Translation of the medieval Latin name used by apothecaries, *morsus diaboli*, which describes the abruptly ending root-stock. According to the 15th century herbal *Ortus Sanitatis*, the Virgin Mary thwarted the Devil's use of the powerful root of *S. pratensis*, whereupon the Devil bit the root in two. See *Scabious*.

DILL (*Anethum graveolens*). Old English *dile*. This and similar names in other Germanic languages may derive from an Indo-european base meaning to blossom.

DITTANDER (*Lepidium latifolium*). XIII century, as a variant of Dittany (for *Dictamnus albus* – see *Dittany* below). *L. latifolium* was grown in medieval and later gardens as a peppery ingredient in sauces, acquiring by the 15th century the names *Dittander* and *Dittany*, the true Dittany (except in a dried state) having been unknown until the late 16th century.

DITTANY (*Dictamnus albus*, native of the Mediterranean region introduced in the 16th century). XIV century. From the Old French *ditan*, from the Latin *dictamnus*, from Greek *diktamnon*, a foreign name folk-etymologically connected by the Ancients with Dikte, the mountain in Crete. *D. albus* was greatly reputed as a wound-herb, since the *diktamnon* with which it was equated was eaten by goats, according to Theophrastus, to expel arrows which had hit them. Crete was long supposed to be the unique source of dittany – 'Dittany of Crete', 'Dittany of Candy'.

DOCK (*Rumex obtusifolius*). Old English *docce*, with cognate names in other Germanic languages. Used in such combinations as *Burdock* (q.v.) for other large-leaved plants.

DODDER (*Cuscuta epithymon, C. europaea*). XIII century. Similar names in German, Swedish, etc. Of uncertain derivation.

DOG-ROSE (*Rosa canina*). XVI century translation of medieval Latin *rosa canina*, translation in turn of Latin (from Greek) *cynorrhodon*. Objurgatory: not a rose of the gardens.

DOG-VIOLET (*Viola canina*). XVIII century. Translation of the botanical Latin *V. canina*. See *Violet*.

DOG'S MERCURY (*Mercurialis perennis*). XVI century. A translation of the herbalists' Latin *mercurialis canina*, the useless Mercury, in contrast to the garden weed *M. annua*, which was used in medicine, and was equated with the *mercurialis* of the Romans, discovered, according to Pliny, by the god Mercury. See also *Mercury*.

DOG'S-TAIL (*Cynosurus cristatus*). XVIII century. A translation of the Linnaean Latin *cynosurus* 1737, modelled on Greek *kuon oura* (describing the shape of the panicle).

DOGWOOD (*Cornus sanguinea*). XVII century. Earlier (XVI century) *Dog-tree, Dog-berry Tree*. Objurgatory, as stated by Parkinson 1640: 'Dogge berry tree, because the berries are not fit to be eaten, or to be given to a dogge.'

DOUGLAS-FIR (*Pseudotsuga taxifolia*, native of the west of North America, introduced 1826). XIX century. Seeds of *P. taxifolia* (formerly *Abies*, or *Pinus, Douglasii*) were sent home from British Columbia in 1826 by the Scottish botanical explorer, David Douglas (1798–1834). (On the same journey he found the now familiar Flowering Currant, *Ribes sanguineum*. In Hawaii in 1834 Douglas and a wild bull fell into a pit-trap. The bull gored him to death.)

DOVE'S-FOOT (*Geranium molle*). XVI century. Also *Culverfoot*, 15th century. Translation of the medieval Latin name *pes columbae* or *pes columbinus*, which in other countries was *G. columbinum* (in England *G. molle* is the commoner species).

DOYENNÉ DU COMICE (variety of dessert pear). XIX century. French 'deanery [pear] of the [horticultural] assembly', *doyenné* being a term for various kinds of pear, shortened from *poire de doyenné*. The Doyenné du Comice was raised 1849–1850 in the Jardin des Beaux Arts at Angers by the gardener M. Dhomme and M. Millet de la Turtandière, president of the *Comice horticole d'Angers*.

DROPWORT (*Filipendula vulgaris*). XV century. The roots have ovoid tubers: 'In the rote are smale pellotis schape as it were rounde pes', Stockholm Medical MS. c. 1425. These 'drops' suggested the virtue ascribed to the plant of opening up urinary strictures, and producing a flow drop by drop.

DRY ROT (the fungus *Serpula lacrymans*; and the effects of its growth on timber). In the second sense, late XVIII century; as name of the fungus, XIX century. A misnomer, since the rotted timber is damp, not dry.

DUKE OF ARGYLL'S TEA-PLANT (*Lycium halimifolium*, native of S.E. Europe and Western Asia, introduced c. 1730). XIX century. At Whitton in Middlesex, Archibald Campbell, 3rd Duke of Argyll, a devotee of trees and shrubs, received a tea plant and a *L. halimifolium*, with their labels, and so their identities, mixed. It seems that the Duke grew the *Lycium* under the wrong label, and then referred to it as his tea-plant.

DULSE (the edible seaweed *Rhodymenia palmata*). XVII century. Adaptation of the Irish name *dūileasg*.

DUSTY MILLER (*Primula auricula*, the Auricula, native in the Alps, introduced towards the close of the 16th century). Early XIX century. The mealy leaves recalled the then familiar sight of a miller white from head to foot.

DUTCH RUSH (*Equisetum hyemale*). XIX century. This Horsetail, not growing very commonly in England, was imported in bundles from Holland as a domestic polisher (for which it is still sold in continental markets).

DUTCHMAN'S BREECHES (*Dicentra cucullaria*, native of North America, introduced 1731). XIX century. Descriptive of the flowers with their widely divergent spurs.

DUTCHMAN'S PIPE (*Aristolochia durior*, native of North America, introduced 1763). XIX century. From the U-shaped tube of the yellowish flower, like the stem of a meerschaum pipe.

DWALE (*Atropa belladonna*, Deadly Nightshade). Not, as one might think, a book name out of poetic fancy. XIII century (= a stupefying drink). XIV century, as the plant which stupefies, by means of its narcotic berries. Cf. German *twalmwurz*, stupefaction plant, 15th century. The Germanic base is a verb meaning 'to delay', 'to send astray'. Cf. also, for *A. belladonna*, the German *dolo*, 11th century; *dolwurz*, 15th century, 'silly plant'.

DYER'S GREENWEED (*Genista tinctoria*). XVI century. (Gerard 1597, *Diers Greening Weede*.) Used in dyeing as a 'greening weed'.

DYER'S ROCKET (*Reseda luteola*). XIX century. Cf. *Dyer's Greenweed*, above. Another greening or yellowing weed of the dyers.

E

EARTH-NUT (tuber of *Conopodium majus*; and the plant. Pig-nut). Old English *eorthhnutu*. xv century, *eorthe-note*. There are similar names in German (*Erdnuss*), Dutch, etc. Cf. French *terre-noix*, and the early botanists' Latin *nucula terrestris*. Children learn to trace the root to the tuber, which they eat.

EARTH-STAR (fungi of the genus *Geastrum*, formerly *Geaster*, especially the star-shaped *G. rufescens*). xix century. Translation of the genetic Latin name *Geaster* (1729) – Greek *ge*, 'earth', + *aster*, 'star'.

EBONY (the timber of *Diospyros ebenum*, native of India). xvi century. Earlier *eben*, *eban*, xv century. From the Old French *ébaine*, from Latin *hebenus*, *ebenus*, from Greek *hebenos* or *ebenos*, from Egyptian *hebni*. See also *Hebenon*.

EDELWEISS (*Leontopodium alpinum*, native of the Pyrenees, the Alps and Central Asiatic mountains). xix century. German *edel*, 'noble' + *weiss*, 'white': 'the noble white one'. Commonly thought to have been coined for 19th century holiday-makers in the Alps; but perhaps a folk-name of the Hohe Tauern range in Austria, which spread through the Alps.

EGG-PLANT (*Solanum melongena* var. *esculentum*, native of Africa and Asia, introduced late in the 16th century). xviii century. The shining white fruit (of the white form) swelling to something of an egg shape.

EGLANTINE (*Rosa rubiginosa*, Sweet-Briar). xiv century. Through French from Provençal *aiglentina*, *aiglent*, from the feminine of a medieval Latin **aculentus*, 'prickly', 'thorny' (Latin *aculeus*, 'a little needle', i.e. spine, prickle).

74

EINKORN (the wheat *Triticum monococcum*, native of Western Asia, which was extensively grown in neolithic agriculture). XX century. German *Einkorn*, 'one grain'.

ELDER (*Sambucus nigra*). Old English *ellern*, *ellen*, *elle* (not infrequent in place-names).

ELECAMPANE (*Inula helenium*, native of Central Asia, introduced in the Middle Ages). XVI century. Adaptation of the medieval apothecaries' Latin name *enula campana*, 'enula of the fields', from Latin *inula*, with which *I. helenium* was equated.

ELTON HEART (variety of cherry). XIX century. Cherry with heart-shaped fruit raised at Elton, Herefordshire, in 1806 by the great horticulturist, Thomas Andrew Knight (1758–1838).

ELM (species of the genus *Ulmus*). Old English *elm*, with cognate words in other West European languages (as Latin *ulmus*, Old Norse *almr*), for trees of a genus which belongs mainly to Western and Northern Europe.

EMMER (species of wheat which was one of the staples of neolithic agriculture, *Triticum dioicum*, Amelcorn, native of Western Asia). XX century. German *emmer*, dialectical shortening of *Amelkorn*, 'starch-grain', *amel* from Latin *amylum*, 'starch'.

ENCHANTER'S NIGHTSHADE (*Circaea lutetiana*). XVI century, Gerard 1597. The Flemish botanist Mathias de l'Obel (1538–1616) equated the Greek plant *kirkaia*, Latin *circaea*, used in charms, first with *Solanum dulcamara*, the Woody Nightshade, then with *C. lutetiana*. *Kirkaia* was taken to mean the plant of the witch or enchantress Kirke, or Circe. So Gerard's 'Enchanter's Nightshade' makes the best of both guesses.

ENDIVE (*Cichorium endivia*, native of Egypt or India, introduced in the 16th century). XV century. Old French *endive*, from medieval Latin *endivia*, *indivia*, from medieval Greek *indivi*, from Greek *entubion*, *entubon*, from Egyptian *tybi*, the month of January, when Endive was eaten. See also *Chicory*.

ESPARTO-GRASS (*Stipa tenacissima*, native of Spain and North Africa). XVIII century. Spanish *esparto*, from the Latin *spartum*, from Greek *sparton*. (Pliny describes at length the collection, treatment and uses of Esparto-grass.)

EUCALYPTUS (species of the genus *Eucalyptus*, Gum-Tree, mostly natives of Australia). XIX century. English use of the botanists' Latin *eucalyptus* (1788), a euphonious name based on Greek *eu-*, 'well', + *kaluptos*, 'covered', describing the cap-like calyx over the unopened flower.

EVENING PRIMROSE (species of the genus *Oenothera*, natives of North and South America, mostly introduced in the 18th and 19th centuries). XIX century. Having flowers which open in the evening. Cf. the German name *Nachtschlüsselblume*, Night Cowslip.

EVERLASTING (species of various genera whose papery dried flowers retain shape and colour; especially *Compositae*, including *Helichrysum orientale*, introduced from South Africa in 1629). XVIII century, shortened from *Everlasting Flower*, translation of the French *fleur immortelle*. Cf. *Eternal Flower*, 1706, for such species of the genus *Xeranthemum*, introduced from Southern Europe. See *Immortelle, Tansy*.

EYEBRIGHT (*Euphrasia officinalis* agg.). XV century, *yeebryght*. The eye-like appearance of the flowers suggested its use in afflictions of the eye, cf. German *ougenklar*, 1500. To apothecaries *E. officinalis* was known as *ocularia* and *ophthalmica*.

F

Fairy Ring Mushroom (*Marasmius oreades*). xix century. Mushroom causing the commoner 'fairy rings' in turf, a name no doubt suggested by Prospero's speech in Shakespeare's *The Tempest* (V, i, 36):

> You demy-Puppets, that
> By Moone-shine doe the greene sowre Ringlets make,
> Whereof the Ewe not bites: and you, whose pastime
> Is to make midnight-mushrumps.

Fat Hen (*Chenopodium album*). xviii century. Properly a name for *Chenopodium bonus-henricus*, Good King Henry (q.v.), eaten as a fatty or greasy spinach-like vegetable; 'hen', ? an abbreviation of Henry or Henricus. The old name *bonus Henricus* translates the German 16th century *Guter Heinrich* – Heinrich as the name of a goblin with a knowledge of healing plants. *C. bonus-henricus* also had its fatty names, *Atriplex unctuosa*, or *Lapathum unctuosum*. (*C. album* has fatty seeds, which were among the early food grains of Europe; they were part of the fertility meal given to Tollund Man, of the 1st century BC, before he was sacrificed in his Danish bog.)

Felwort (*Gentianella amarella*). Old English *feldwyrt*, 'field plant', which was the Gentian of medicine, the tall bitter *Gentiana lutea* of the Alps, etc.

Fennel (*Foeniculum vulgare*). Old English *finugl*, via medieval Latin *fenuclum*, from Latin *faeniculum*, a diminutive of *faenum*, hay: 'little hay' – little in reference to the finely cut leaves. For 'hay' cf. the use of 'grass' in English names of several ungrasslike plants.

77

FENUGREEK (*Trigonella foenum-graecum*, native of Western Asia). XIV century. From the Old French *fenugrec*, from Latin *faenum Graecum*, 'Greek hay'.

FERN (plant of the class *Filicinae*). Old English *fearn*, with cognate names in other Germanic languages, from a presumed Indoeuropean word with the meaning of 'frond or leaf like a feather'.

FESCUE (grasses of the genus *Festuca*). In this sense XVIII century, as a translation of the Linnaean generic name *Festuca*, from Latin *festuca*, 'a straw', or 'wild oats'. (*Fescue*, for a straw, a mere nothing, is found in the 14th century.)

FEVERFEW (*Chrysanthemum parthenium*, native of S.E. Europe and Western Asia, introduced in the Middle Ages). XV century. From the Anglo-Norman **fevrefue*, from the medieval Latin *febrifugia*, 'that which puts fever to flight'; one of the universal plants of medicine, aromatic and cheering, used much as we use aspirin, against fever, headache, ague, etc. The Old English name was *feferfuge*.

FIG (fruit of *Ficus carica*; and the tree, native of the Mediterranean region, introduced in the 16th century). XIII century. From the Old French *fige*, from the Provençal *figo*, from the Low Latin *fica*, Latin *ficus*.

FIGWORT (*Scrophularia nodosa*). XVI century, first applied (Turner 1548) to *Ranunculus ficaria*, Lesser Celandine; then (Lyte 1578) to *S. nodosa*. The Romans likened the swellings of piles to figs, so that *ficus* meant the disease as well as the fruit. In medieval medicine *ficus*, or the 'fig', demanded the application of *ficariae*, plants whose virtue against the disease was indicated by root swellings (*S. nodosa*) or tubers (*R. ficaria*). *S. nodosa* was the *ficaria major*; *R. ficaria* was the *ficaria minor*.

FILBERT (fruit of the cultivated Hazelnut-tree, *Corylus maxima*, native of S.E. Europe and Western Asia). XIV century, *philliberd*. From the Normandy name *noix de filbert*, nut of (St) Phillibert, abbot of Jumièges in Normandy (d. 684), since the nuts mature at the time of his feast day, towards the end of August.

FIORIN (the grass *Agrostis stolonifera*). XIX century. Irish *fiorthann*; a name, 'Irish Fiorin or Fyoreen Grass', popularized by the Irish agricultural cleric William Richardson (1740–1820), who wrote extensively on the agricultural benefits which would follow from the wide planting of *A. stolonifera*, especially in Ireland.

FIR (*Pinus sylvestris*, etc.). Old English *furh, fyrh*; but not commonly used again until the 14th century. There are similar names in other Germanic languages.

FIREWEED (*Chamaenerion angustifolium*). ? XIX century in this sense, in America; perhaps as a translation of the German name *Feuerkraut*, noted for this plant by Gesner in 1561, with the explanation that it flourished (as it does) on woodland ground cleared by fire.

FLAG (*Iris pseudacorus*). A name of obscure origin and affinities, not found until the XIV century; used for marsh plants – sedges, reeds, rushes and iris.

FLAX (*Linum usitatissimum*). Old English *fleax*, with cognate West Germanic names, from an Indoeuropean *plek*, to twist or braid or interweave.

FLEABANE (*Pulicaria dysenterica*). XVI century, Turner 1548. Taken to be the *konuza* of Dioscorides, which warded off fleas and midges, and was used by the Romans to make chaplets.

FLIXWEED (*Descurania sophia*). XVI century. The seed of this once common English plant was taken against the flix or flux, i.e. dysentery. 'The seede of Flixweede or Sophia broken with wine or water of the Smithes forge, stoppeth the bloudy flixe, the laske, and all other issue of bloud' (Lyte 1578).

FLORA (as *1.* a catalogue of the species of flowering plants in a district, county or country; *2.* a term for the plant life of a district or a period of time, a 'rich flora', a 'scanty flora', etc.).

XVIII century. From Flora, the Italian goddess of the blossoming of flowers, 'whose festival was celebrated on the 28th of April, often with unbridled licence' (Lewis and Short's *Latin Dictionary*), from Latin *flos*, genitive *floris*, 'a flower'. The *Oxford English Dictionary* finds its first quotation for sense 2 appropriately in Gilbert White's *Natural History of Selborne* (1789) in a letter dated 1778. The first example of *flora* in sense 1 is quoted as John Lightfoot's *Flora Scotica*, published in 1777.

FLOTE-GRASS (*Glyceria fluitans*). XV century. *flote*, i.e. *float*, from the floating leaves.

FLOUR (finely ground cereal grains). XIII century. The 'flower', or finer and better part of meal (q.v.), *flour* preserving the medieval spelling of *flower*, from Old French *flour*, from Latin *flos*, genitive *floris*.

FLUELLEN (*Veronica officinalis*). XVI century, Turner 1548. In Lyte 1578, *Herbe-fluellyn*, which is nearer the Welsh from which the name is adapted, *llysiau Llywelyn*, 'herbs of (St) Llywelyn', i.e. herbs which are in flower around 7 April, which is the feast day shared by St Llywelyn and his son St Gwrneth. The two are always spoken of together, and *Gwrneth*, literally 'great strength', is another of the Welsh names for *V. officinalis*, as a medicinal plant.

FOG (grass of the aftermath, coarse winter grass, chiefly North Country dialect; also, in the originally local name, Yorkshire Fog, for the common grass *Holcus lanatus*). XIV century. ? from Old Norse **fogg*, 'long, lax, damp grass'.

FOOL'S PARSLEY (*Aethusa cynapium*). XVIII century. Translation of the *apium rusticum* of 16th century botanists and apothecaries. Cf. French *persil des fous*, German *Narrenpetersilie*.

FORGET-ME-NOT (species of the genus *Myosotis*, especially *M. palustris*). XVI century. A translation of the French *Ne m'oubliez mye*, which in turn was a translation of the German *Vergissmeinnicht* (16th century) as a blue flower worn to retain a lover's affection. It was first positively given as a name for

Myosotis by the Swiss polymath Konrad Gesner in 1561. *Forget-me-not* became the universal *Myosotis* name in English only in the 19th century, Coleridge having published in the *Morning Post* in 1802 his poem 'The Keepsake', in which he wrote of *That blue and bright-eyed flowerlet of the brook,/Hope's gentle gem, the sweet Forget-me-not*, with a note to say that the German equivalent was used 'over the whole Empire of Germany'.

FORSYTHIA (shrubs of the genus *Forsythia*, introduced from China, 1845). XIX century. English use of the generic Latin name honouring William Forsyth (1737–1804), superintendent of the royal gardens at Kensington Palace and St James's Palace.

FOXGLOVE (*Digitalis purpurea*). Old English *foxes glōfa*, 'fox's glove', but first for the Deadly Nightshade, *Atropa belladonna*; by the XIV century the name for *D. purpurea*. Names in other languages give no help in explaining a special association between foxes and the finger-stall shaped flowers. The local Deadly Nightshade and the common Foxglove, however, are both woodland plants, which can often be seen on disturbed ground by a fox's earth: that the flowers could be gloves worn by a fox is an easy enough fancy.

FOXTAIL (the grass *Alopecurus pratensis*). XVI century (though not at first for *A. pratensis*) translating botanists' Latin *cauda vulpina*, or *gramen alopecuroides*, Theophrastus (and Pliny) having written of a grass *alopekouros* (Greek *alopex*, 'fox', *oura*, 'tail') with a soft panicle like a fox's brush.

FRANGIPANE (the tree *Plumeria rubra*, Red Jasmine, native of Mexico and northern South America; and the scent derived from its flowers). In the second sense XVII century; in the first sense XIX century. Not from some language of the scented East, but from French *frangipane* (1588), a perfume for gloves devised by the Italian nobleman Marchese Muzio Frangipani.

FREESIA (species of the genus *Freesia*, natives of South Africa). XIX century. Generic name in honour of a Kiel doctor F.H.T. Freese, who died in 1876.

FRENCH BEAN (*Phaseolus vulgaris*, native of the New World, introduced in the 16th century). XVII century. *French*, from being commonly grown in France.

FRITILLARY (*Fritillaria meleagris*, native of Europe, introduced in the 16th century). XVII century. Adapted from the botanists' Latin (1570) *fritillaria*, coined from Latin *fritillus*, a dice-box, as if a box of the kind which holds dice and pieces and opens into a draughtboard or chessboard. Charles de l'Ecluse in his book on the plants of Austria and Hungary (1583) says that the name *fritillaria* was invented for *F. meleagris* by an apothecary at Orléans: the perianth is chequered in its purple tones.

FROG-BIT (*Hydrocharis morsus-ranae*). XVI century, Lyte 1578. Translation of the 16th century botanists' name *morsus ranae*, as if a food for frogs, or as if frogs had bitten away leaves to form the stipules.

FUCHSIA (species of the genus *Fuchsia*, mostly natives of Mexico, Central America and South America, and first extensively grown in the early Victorian period). XIX century. The generic name given by Charles Plumier in 1703, after the discovery of the first known species, to honour the memory of the great German botanist Leonard Fuchs (1501–1566).

FUMITORY (*Fumaria officinalis*). XV century, *fumeterre*, from the Old French *fumeterre*, from the apothecaries' medieval Latin *fumus terrae*, 'smoke of the earth', descriptive of the way *F. officinalis* spreads its pallid blue-green smoke of leaves across garden ground, and from the nitrous fume of its roots.

FUNGUS (especially a mushroom, toadstool, in the huge vegetable group of *Fungi*). XVI century. Latin *fungus*, 'a mushroom', 'something soft and spongy'; used in Latin comedy as an insult, as we might say 'softy' (which suggests that the kinds which the name *fungus* brought to the Roman mind were their favourite *boleti*). The Greek *spoggos*, *sphoggus*, sponge, is related.

FURBELOWS (the large seaweed *Saccorhiza polyschides*). XIX century. Describing the wavy, flounced or furbelowed stipe.

FURZE (*Ulex europaeus* and *U. gallii*). Old English *fyrs*, which may be related to Old English *fyrh, furh*, 'fir (-tree)'. See also *Whin*.

G

GAGE. See *Greengage*.

GALE (*Myrica gale*, Bog Myrtle, Sweet Gale). Old English *gagel* – as in modern Dutch and German – of unknown significance. Often in Old English place-names, e.g. the Devonshire farm name Galsworthy, 'gale enclosure'.

GALINGALE (*Cyperus longus*). XVI century in this sense, Lyte 1578 'English galangal'. Old English *gallengar*, dried roots of *Alpinia officinarum*, of the ginger family, from South China, medieval Latin *gallingar*. The name reached Europe from the Chinese *kao-liang-kiang*, 'ginger of Kao-liang'. The rhizome of the Wild, or English, Galingale was a substitute.

GALL (growth on plants caused chiefly by insects and mites). XIV century. From the Old French *galle*, from Latin *galla*, oak apple.

GALLANT SOLDIER (*Galinsoga parviflora*, Kew Weed, native of Peru, introduced to Kew Gardens 1796, observed outside the gardens by 1860). A garbling of the generic name, which commemorates the 18th century Madrid botanist M. M. de Galinsoga.

GAMBOGE (resin, giving the yellow pigment anciently used in Chinese painting and imported to the West since the 16th century, of the tree *Garcinia hanburyi*, native of Cambodia and Siam). XVII century, *cambugia*; XVIII century, *gamboge*. From the apothecaries' Latin *gambogium*, 'Cambodian (substance)', i.e. from Kambuja.

GARDENIA (species, and their flowers, of the genus *Gardenia*, especially *G. jasminoides*, native of China, the first kind to be introduced, in 1735). XIX century. English use of the Latinized

84

11. Galingale (Fuchs, *De Historia Stirpium*, 1542)

generic name given in honour of the Anglo-American botanist Alexander Garden (1730–1791).

GARLIC (*Allium sativum*). Old English *gārleāc, gār*, 'spear(head)' + *leāc*, 'leek', i.e. the leek which has cloves like spearheads. 'Take nine *clufa garleaces*,' says a prescription in the *Lacnunga* of c. 1000 AD. See *Leek*.

GARLIC MUSTARD (*Alliaria petiolata*, Jack-by-the-Hedge). XVI century (Gerard 1597).

GEAN (*Prunus avium*, the wild Sweet Cherry; the nurseryman's name for rootstock of *P. avium* for grafting). XVI century, from the Old French *guine*.

GENTIAN (species of the genus *Gentiana*). XIV century. From the Latin *gentiana*, from Greek *gentiane*. Referring to the tall coarse *G. lutea* of the Alps, the Gentian whose root is the source of medicinal bitters and the flavouring of liqueurs, Pliny wrote that it was discovered (as a medicinal herb) by 'Gentius, a king of the Illyrians' – a sample of ancient folk-etymology.

GERANIUM (now usually for species of the genus *Pelargonium*, natives of South Africa – family *Geraniaceae* – and their garden forms. Many species were introduced, as one kind and another of *Geranium Africanum*, or *African Geranium*, in the 18th century; many more in the early decades of the 19th century). In this sense XVII century; but XVI century for native species of the genus *Geranium*, the Cranesbills. From Latin *geranium*, from Greek *geranion*, from *geranos*, 'a crane', in reference to characteristically long and noticeable beak of the carpel. See *Pelargonium*.

GERMANDER *1*. (*Teucrium chamaedrys*, Wall Germander, native of Central and Southern Europe, introduced as a medicinal plant). XV century. From the medieval apothecaries' Latin *gamandrea*, which derives from the Greek *khamaidrus, khamai*, 'on the ground', + *drus*, 'oak', 'ground oak'. 'Chamaedrys is a plant with mint-sized leaves, like oak leaves in colour and indentation' (Pliny).

2. (*Veronica chamaedrys*). XVI century, *Germander, English*

86

Germander, Wild Germander (Lyte 1578), the leaves having some resemblance to those of *Teucrium chamaedrys*. Later, 19th century, distinguished as *Germander Speedwell*.

GHERKIN (small pickling or pickled fruit of the Cucumber, *Cucumis sativus*). XVII century. From the plural of the Dutch *gurk*, cucumber, which, through German *gurke*, Polish *ogurek* (similar forms in other Slavonic languages), derives from late Greek *aggourion*, a diminutive from *agourous*, 'not ripe', 'green': i.e. a something which is small and immature. See *Cucumber*.

GILLYFLOWER (*Dianthus caryophyllus*, Clove-Scented Pink; *Cheiranthus cheiri*, Wallflower, etc.). XV century, as a name for various flowers, from their scent of *girofle* or Cloves (q.v.). See also *Stock*.

GIN (drink distilled from grain and originally flavoured with juniper berries). XVIII century. Shortened from *geneva*, assimilated to the spelling of the Swiss city, from Dutch *genever*, from Old French *genevre*, from Latin *juniperus*, 'juniper'.

GINGER (rootstock of *Zingiber officinale*, native of the islands of the Pacific). XIII century. One of those words which have wandered through time, space and language, attached to the spread round the world of the substance they describe. Its backward course leads through French, Latin (*zingiber*), Greek, Pali, Middle Indian, to a Dravidian or earlier origin. The meaning is 'horn-root', root with a horn shape.

GINGKO (*Gingko biloba*, Maidenhair Tree, native of China, introduced from Japan 1754). XVIII century. From Sino-Japanese *ginkyō*, from the Old Chinese *ngin-ghang* meaning 'silver apricot', whence also Peking Mandarin *yin-hsing*. (The usual Japanese name for the Gingko is *ichō*.)

GIPSYWORT (*Lycopus europaeus*). XVIII century, earlier *Gipsy-herb*; *Egyptian Herb*, which is Lyte's translation, 1578, of Flemish *Egyptenaers cruyt*, 'that is to say, the Egyptians herbe, bycause of the Rogues and runnegates which call themselves Egyptians, do

87

colour themselves blacke with this herbe'. *L. europaeus* gives a blackish dye.

GLADDON (*Iris foetidissima*). Old English *glædene*. The name has now been restricted by books to *I. foetidissima*, but was earlier a common name for the Yellow Iris, *I. pseudacorus*. It is in some way derived from Latin *gladiolus* (which in Pliny is the Field Gladiolus, *G. segetum*, of cornfields, etc., in the Mediterranean region), 'little sword', diminutive of *gladius*, 'a sword', describing the leaf-shape. The Yellow Iris was known to early botanists as *gladiolus, gladiolus aquaticus*, 'water gladiolus'. Lyte 1578 distinguished *I. foetidissima* as 'stinking Gladyn'.

GLADIOLUS (commonly the garden kinds of the genus *Gladiolus*, mostly descended from South African species introduced from mid 18th century onwards). XVI century (for *Iris pseudacorus*); XVII century for *Gladiolus*. For derivation see *Gladdon* above.

GLASSWORT (species of the marshland genus *Salicornia*). XVI century, Gerard 1597. Glasswort was reduced to ashes providing glassmakers with alkali (carbonate of soda).

GLASTONBURY THORN (nurserymen sell an early flowering form of *Crataegus*, which they call *praecox*, as 'Glastonbury Thorn'). XVII century, in John Aubrey's *Natural History of Wiltshire*, 1656–1675. (The anonymous metrical *Lyfe of Joseph of Armathia* 1520 mentions 'thre hawthornes', growing on the hill Werall, at Glastonbury, which

> Do burge and bere grene leaves at Christmas
> As freshe as other in May,

but does not say that a Christmas-flowering thorn on the hill had grown from the staff of Joseph of Arimathea, a legend recorded by the Roman Catholic antiquary Charles Eyston (1667–1721), in an account of Glastonbury written in 1716 and included in Thomas Hearne's *History and Antiquities of Glastonbury*, 1722.)

GLOBE-FLOWER (*Trollius europaeus*). XVI century, Gerard 1597, from the shape of the flowers.

GOAT'S-BEARD (*Tragopogon pratensis*). XVI century, Turner 1548. Translation of the herbalists' Latin *barba hirci*, translating in turn the Greek *tragopogon*, Salsify (*T. porrifolius*), which has similar long white pappus not unlike a billygoat's beard.

GOAT TANG (the seaweed *Polyides caprinus*). See *Tang*.

GOLD (*Chrysanthemum segetum*, Corn Marigold). Old English *golde*, from *gold* (the metal), i.e. golden-coloured. The name is found in many place-names (though in names to do with streams, fords, bridges, etc., *golde* signifies Marsh Marigold, *Caltha palustris*). Gold remained the common name through the centuries, every farmer on light soil being too well aware of the golden flowers of a plant which was one of the worst of weeds.

GOLD OF PLEASURE (*Camelina sativa*). XVI century, Gerard 1597. ? coined for a Gold (see *Gold* above) which is useful or affords pleasure, since this yellow-flowered plant, though often a weed, was grown abroad for its yellow seeds and the oil they gave.

GOLDEN-ROD (*Solidago virgaurea*). XVI century, Turner 1568, translating the botanists' Latin *virga aurea*.

GOOD KING HENRY (*Chenopodium bonus-henricus*). XVI century. Lyte 1578 calls *C. bonus-henricus* 'Good Henry', translating the German *Guter Heinrich*. 'King' was soon interpolated (Gerard 1597). *C. bonus-henricus* was taken to be a kind of Mercury; and as an old pot-herb of good medicinal reputation, the *Tota Bona*, 'Allgood', of apothecaries, it was distinguished as the Good Henry from *Mercurialis annua*, the garden weed, the *Böser Heinrich*, or Bad Henry.

GOOSEBERRY (*Ribes uva-crispa*, the fruit and the plant). XVI century. Frequently assumed to be *goose* + *berry*, but probably a garbling, assimilated to *goose*, either of French *groseille* (= Red Currant), or of the Latinization of *groseille* by 16th century botanists as *grossula*. The fruit in English are also *grosers, grosiers, grozers*, apparently adaptations of *groseille*. Turner 1548 says, 'Uva crispa is also called Grossularia, in english a Groser bushe,

89

a Goosebery bush.' Lyte 1578 follows his chapter on Goose-berries ('in Frenche, *des Groisselles*') with a chapter describing Red and Black Gooseberries, i.e. Red and Black Currants: 'The first kinde is called *Grossulae rubrae*, *Ribes rubrum*: in Englishe, Redde Gooseberies, Beyondsea Gooseberies . . . in Frenche, *Groiselles rouges* . . . The second kinde is called *Ribes nigrum*: in English, Blacke Gooseberies or blacke Ribes: in Frenche, *groiselles noires.*'

Groiseille is from the Old French *grozelle*, from the Frankish *krûsil*, cognate with German *kraus*, 'crisp'.

GOOSEFOOT (*Chenopodium album*, Fat Hen, and other species of *Chenopodium*). XVI century, coined by Turner 1548 for the botanists' Latin *pes anserinus*, or German *Gänsefuss*.

GOOSEGRASS (*Galium aparine*, Cleavers). XVI century, Turner 1538 (in the XV century for *Potentilla anserina*, Silver-weed). *G. aparine* is eaten by geese, and is, or was, fed to goslings. 'he wol make gees and hennes fatte if this herbe be broke smale among hure met.' MS. Harley 3840, British Museum (15th century), in *Agnus Castus*, ed. G. Brodin, 1950.

GOPHERWOOD (wood from which the Ark was made, ? timber of the cypress *Cupressus sempervirens*, from Western Asia). XVII century, Authorized Version of the Bible 1611 (*Genesis* vi,14). Hebrew *gōpher*, from which Greek *kuparissos*, English *cypress* may derive. See *Cypress*.

GOURD (ornamental hard-shelled fruits of various species of *Lagenaria*, *Cucurbita*, *Cucumis*, etc.). XIV century. From the Latin *cucurbita* (Pliny has much to say of gourds), via French *gourde* (*gouhorde*, *cougorde*), and Anglo-Norman *gurde*. At first for the bottle-shaped gourds of *Lagenaria siceraria*.

GOUTWEED (*Aegopodium podagraria*, Ground Elder). XVIII century. Earlier (Gerard 1597) *Goutwort*, translating the botanists' Latin *herba podagraria* (Latin *podagra*, from Greek *podagra*, 'gout'). The leaves, 'pounded and layde uppon suche members or partes of the body, as are troubled and vexed with

12. Gourd (Fuchs, *De Historia Stirpium*, 1542)

the gowte, swageth the payne, and taketh away the Swelling'
(Lyte 1578).

GOWAN (*Bellis perennis*, the Daisy, also formerly for other plants
with yellow-centred, or yellow flowers). XVI century. A variant
of *gollan, golland* for buttercups, Globe-flower, Marsh-marigold
(14th century), meaning 'a yellow flower' (cf. Norwegian *gal,
gaul*, yellow).

GRAM (*Phaseolus mungo*, Black Gram, with black seeds; *Phaseolus
aureus*, Green Gram, Mung, with green seeds; both probably of
East Indian origin). XIX century. From Portuguese *graõ*, from
Latin *granum*, 'a seed' or 'grain'.

GRANADILLA (*Passiflora edulis*, native of Brazil, and *P. quad-
rangularis*, from tropical America, the Passion-flowers, and their
fruits). XVII century. Spanish *granadilla*, 'little pomegranate'
(*granada*). See *Pomegranate*.

GRAPE (fruit of the vine). XIII century. From Old French
grape, grappe, crape, 'bunch of grapes', i.e. that which was cut off,
at the *vendange* – before the modern introduction of secateurs –
with a *grape*, a small bill-hook (such as may be seen in medieval
calendar illuminations or woodcuts for September, the grape-
harvest month), the word *grape, crape* having a German origin,
cf. Old High German *krapfo*, 'a hook'.

GRAPEFRUIT (fruit of *Citrus paradisi*, which arose apparently
as a sport or by hybridization in Barbados before 1809). XIX
century. The fruit grow in clusters, as if – rather a forced resemblance – bunches of outsize grapes.

GRAPE-HYACINTH (*Muscari racemosa*, etc.). XIX century,
translating the (16th century) botanists' Latin *hyacinthus
botryoides*, 'hyacinth like a bunch of grapes' (Greek *botrus*), from
the packed terminal raceme of flowers.

GREATER CELANDINE. See *Celandine*.

GRASS (collectively). Old English *gærs, græs, gres*, meaning at
first all green stuff eaten by cattle, and going back, with cognate

words in the other Germanic languages, to a Germanic base *gro-*, 'to grow'. From this base descend *grow, green, grass,* and *graze*.

GRASS OF PARNASSUS (*Parnassia palustris*). XVI century, Lyte 1578, translating the botanists' Latin *gramen Parnasium*, translating in turn the Greek *agrostis en Parnasso*, mentioned by Dioscorides, the agrostis growing on Parnassus, the mountain of Apollo and the Muses, which was taken to be *P. palustris*.

GRAVENSTEIN (variety of apple). XIX century. From the Danish apple-growing village of Gravenstein, near Sonderburg, close to the Flensburger Fjord, where the variety was found in the orchard of the castle.

GREAT BURNET (*Sanguisorba officinalis*). XVI century. See *Burnet*.

GREENGAGE (related kinds of a green-fruited variety of plum, *Prunus domestica* ssp. *italica*; and the fruit; ? from Asia Minor). XVIII century. Named c. 1725 after Sir William Gage (1657–1727) of Hengrave Hall, Bury St Edmunds, Suffolk, who brought back plum-trees of this group from France in 1724; the label was lost, so a name was required. See *Reine Claude*.

GREENWEED. See *Dyer's Greenweed*.

GRIM THE COLLIER (*Hieracium brunneocroceum*, native of Central Europe, introduced in the 17th century). XVII century (Parkinson 1629, a garden name 'both idle and foolish'). From the dusky, coalman-like appearance of the flowers. Cf. the play by William Haughton (c. 1575–1605), *Grim the Collier of Croyden; or, The Devil and his Dame*, c. 1600.

GRISETTE (the edible mushroom *Amanita vaginata*). XX century. The French name *Grisette*, diminutive of *gris*, grey, 'little grey one', on account of the mouse-grey cap and stem.

GROMWELL (*Lithospermum officinale*). XIV century. From the Old French *gromil*, of which the second syllable -*mil* is from the Latin *milium*, 'millet'. In medieval Latin *L. officinale*, a plant with a surprisingly strong medicinal reputation, was named *milium solis*, 'millet of the sun', on account of its hard, glistening seeds. The first syllable of the name has not been convincingly explained.

GROUND-IVY (*Glechoma hederacea*). XIV century, *ivy of grownde*. Translation of medieval Latin *edera terrestris* (as distinct from *Edera arborea*, 'tree-ivy'), which in turn translated the *chamaicissos* of Pliny and *khamaikissos* of Dioscorides, 'ivy on the ground'.

GROUND-PINE (*Ajuga chamaepitys*). XVI century, Turner 1548, translating the *chamaepitys* of Pliny, Greek 'pine on the ground', describing the delightful resinous smell when crushed: 'of the savour of the Pyne, or Fyrre tree' (Lyte 1578).

GROUNDSEL (*Senecio vulgaris*). Old English *grundeswylige*, 'ground-swallow(er)', an apt name for a weed of habits familiar to every gardener; but it also has the earlier, 7th century name *gundesuilge*, which should mean 'swallower of matter' (*gund*, in later English *gound*, 'matter discharged from the eyes'). There is no later occurrence of this medicinal *gundesuilge*, nor is there any equivalent name, which suggests that *gunde* was a misspelling of *grunde* in the manuscript glossaries in which it appears. However, groundsel was administered for 'bleared and dropping eyes' (Lyte 1578), and for 'inflammations or watering of the Eyes by reason of the Defluxion of Rheum into them' (Culpeper 1653), and was used as a poulticing herb in the Anglo-Saxon period.

GUAVA (species of the Central and South America genus *Psidium*; and the fruit). XVI century, from Spanish *guayaba*, adaptation of the name learnt by Columbus and his men from the Arawak Indians of the West Indies.

GUELDER ROSE (*Viburnum opulus* var. *roseum*, Snowball Tree, introduced from Holland in the 16th century). XVI century, Gerard 1597, *Gelders Rose*. From Dutch *Geldersche roos*, i.e. the tree, with rose-like bloom, from the Dutch province of Guelders, the modern Gelderland.

GUERNSEY LILY (*Nerine sarniensis*, native of South Africa, introduced in 1659). XVII century, Evelyn 1664. The plant was naturalized on Guernsey, where it was said (1680) to have grown from bulbs washed out of the wreck of a ship from Japan. Evelyn also called it *Narcissus of Japan*: it did in fact grow in Japan in the 17th century.

GUM-TREE (species of the genus *Eucalyptus*, chiefly natives of Australia). XIX century, from the exudation of gum-like resin. (XVII century in North America for trees of the genus *Nyssa*.)

GUMBO (*Hibiscus esculentus*, native of tropical Africa, and its edible fruits; Okra, Lady's Fingers). XIX century, from the American French *gombo*, by origin a West African name, ? *ochinggombo*, by which *H. esculentus* is known in the Angola language Umbundu. See *Okra*.

GUM TRAGACANTH (exudation from the roots of *Astragulus gummifer*, native of the Near East). XVI century. From French *tragacanthe*, from Latin *tragacantha*, from Greek *tragakantha*, 'goat thorn', the name of the plant.

GUNNERA (species of the genus *Gunnera*, natives of South America, etc., Prickly Rhubarb). XIX century. English use of the generic name given by Linnaeus in honour of the Norwegian botanist Johan Ernst Gunnerus (1718–1783), Bishop of Trondhjem.

H

HARDHEADS (*Centaurea nigra*). XVIII century. Descriptive of the hard compact flowerheads. Cf. German *Hardkopp*, Dutch *hardekopp*.

HARDOKES, HOR-DOCKS (? *Arctium lappa*, Burdock). XVII century. Occurs only in Shakespeare's *King Lear*, Act IV, opening of Scene iv, in Cordelia's description of her mad father crowned with weeds. Editors have often changed it to *burdock*, *burdocks*. Though it could well be Old English *hār*, 'grey', 'hoar', + *docce* (cf. Old English *hārhune*, horehound), the first element seems likelier to be 'hair', Old English *hær*, cf. German names for *A. lappa*, e.g. *hoortockers*, 'hair balls', from the burs which children throw at each other so that they stick in the hair.

HAREBELL (*Campanula rotundifolia*). XIV century, but without certainty as to the species. For *C. rotundifolia* the name *Harebell* is not recorded until the XVIII century. Earlier, in the 16th century (Gerard 1597, *Hyacinthus anglicus*, 'Blew English Hare-bels'), it is used for *Endymion nonscriptus*, the Bluebell; which is the sense of 'the azur'd harebell like thy veins', in Shakespeare's *Cymbeline*. The derivation is *hare* + *bell* (not *hair* + *bell*, in spite of the hair-like pedicels of *C. rotundifolia*). The vernacular names of *C. rotundifolia* suggest a connection with the hare as a magic animal; and suggest also that the name *Harebell* was transferred from *C. rotundifolia* to *E. nonscriptus*.

HARE'S-EAR (*Bupleurum rotundifolium*, Thorow-wax). XVI century, Gerard 1597, a description of the 'hollow' ear-shaped leaf, as Gerard says, translating the German *Hasenöhrlein* or the *Auricula Leporis* of the 16th century herbalists.

96

HARE'S-FOOT (*Trifolium arvense*). XVI century, Turner 1562, translating the 16th century botanists' *lagopus*, from Pliny, and the Greek *lagopous*.

HARICOT BEANS (dried beans of *Phaseolus vulgaris*, Kidney-bean, French Bean). XVII century. From the French *haricot*, of disputed origin, commonly referred to the ragout of mutton and vegetables, the *haricot*, as an ingredient; but probably a garbling of *ayacotli*, in the Nahuatl language of the Aztecs.

HART'S-TONGUE (the fern *Phyllitis scolopendrium*). XIV century. Translation of the apothecaries' medieval Latin *lingua cervi*.

HASHISH (dried leaves of the drug plant *Cannabis sativa*, Hemp, native of temperate Asia). XVI century; *hash*, XX century. From Arabic *hashīsh*, 'hay', 'dry herb' (cf. *assassin*, from Arabic *hashshāshīn*, 'hashish-eaters'). See also *Hemp*.

HASSOCK (a clump of grass, sedge or reed in a boggy area). Old English *hassuc*, one of the ancient plant-names with the suffix *-uc*, *-oc*. (Hence also a kneeler, as in church, made originally with reed or sedge.)

HAUTBOIS, HAUTBOY (*Fragaria moschata*, native of Central and S.E. Europe, introduced, probably in the 17th century). XVIII century (Miller 1731). French *haut* + *bois*, 'high wood (strawberry)', *F. moschata* lifting its fruit above the leaves in contrast to the common Wood Strawberry (*F. vesca*) with fruits on the ground.

HAWKWEED (species of the genus *Hieracium*). XVI century, Turner 1562, translating the *hieracion* of Pliny, Greek *hierakion*, from *hierax*, 'a hawk'. Pliny wrote that the *hieracion* of the Ancients (not one of the hawkweeds) acquired its name 'because hawks tear it apart and wet their eyes with the juice, so dispelling dimness of sight, when it comes on them'.

HAW(s) (fruit[s] of the Hawthorn). Old English *haga*, probably shortened from **hagu-berige*, 'hedge-berry'.

HAWTHORN (*Crataegus oxyacanthoides* and *C. monogya*, White-thorn, May-tree). Old English *hagu-thorn*; *haga*, 'haw', + *thorn*.

HAY (cut grass, mowing grass). Old English *hēg*, *hīeg*. Cognate words in other Germanic languages have the basic meaning of 'that which is cut'. Cf. the old English *hēawan*, 'to cut'.

HAZEL (*Corylus avellana*). Old English *hæsel*, *hesel*, in its various forms common to the Germanic languages. The Indoeuropean base from which it descends is ancestral also to the *corylus* of the Romans.

HEADACHE (*Papaver rhoeas*, the Poppy). A common name in the English counties, used for instance by John Clare in his poems. From the employment of poppies against headache and migraine. Cf., from a 15th century herbal in the Bodleian Library, Oxford, 'also yff a man have the mygreme or hed-ache poyn (prick) thys herbe and temper hit with aysell (vinegar) and make a plaster and ley to the fore-hede and to the templys et sessabit [and it will cease]'.

HEARTSEASE (*Viola tricolor* var. *hortensis*). XVI century, used of garden pansies, the Wild Pansy, and the Wallflower (*Cheiranthus cheiri*). ? giving ease of heart by the perfume of the flower. *Viola tricolor* also symbolized love, and easing of the heart, in the arrangement of the petals, cf. the name for Wild Pansy recorded by Turner 1548, *Two faces in a hoode*.

HEATH (*Calluna vulgaris* and species of the genus *Erica*). Old English *hæth*, both untilled tract of country and that which grows on such land and distinguishes it. A word in various forms common to Germanic languages.

HEATHER (also species of the genus *Erica* and *Calluna vulgaris*). Old English **hæddre*, in place-names, otherwise on record first in the XIV century as a Scottish word, *hathir* (later *hedder*). Influenced by *heath*, the spelling *heather* begins in the 18th century. Presumably cognate with *heath*.

HEBENON, HEBONA, HEBON (plant whose deadly juice, in Shakespeare's *Hamlet*, I, v, 62, was poured into the ear of Hamlet's father). What Shakespeare meant by 'cursed Hebenon' (First Folio) or 'cursed Hebona' (the Quartos) has been debated without a sense of how poets make use of words. Commentators have pursued a particular plant which has particular deadly effects, instead of a plant-word of death. Without caring what it signified exactly, Shakespeare must have picked up his juice of Hebenon or Hebona from the 'iouyce of *Hebon*' in Marlowe's *The Jew of Malta*, written about 1590. Barabbas hopes that the poisoned rice pottage he is sending his daughter will work like the Borgias' wine and the blood of the Hydra and the 'iouyce of Hebon' and the poisons of the Styx and the breath of its tributary the Cocytus. Marlowe and Shakespeare would both have known the Garden of Proserpina in Spenser's *Faerie Queene* (II, vii, stanzas 51–6), encircled by the Cocytus, in which the trees and plants

> direful deadly blacke both leafe and bloom
> Fit to adorne the dead, and decke the drery toombe

included the 'Heben sad' and the 'mournfull Cypresse', and

> Dead sleeping *Poppy*, the blacke *Hellebore*,
> Cold Coloquintida, and *Tetra* mad,
> Mortall *Samnitis*, and *Cicuta* bad.

Spenser in turn would have known Gower's 'hebenus that slepy tre' which furnished the boards of the couch of Sleep (*Confessio Amantis*, II, 103). So Hebenon, Hebona simply derives from Latin *hebenus*, *ebenus*, the black-timbered and therefore funereal ebony tree. See *Ebony*.

HELIOTROPE (species of the genus *Heliotropium*, especially *H. europaeum* from S.E. Europe, introduced in the 16th century, and *H. arborescens*, Cherry Pie, introduced from Peru in 1757). XVII century. From French *héliotrope*, from the Latin *heliotropium*, Greek *heliotropion*; *helios*, 'sun', + *trepein*, 'to turn' – since the one-sided flower spikes of *H. europaeum* turn with the sun.

HELLEBORE (species of the genus *Helleborus*). XIV century. From Latin *elleborus*, Greek *helleboros*.

HELLEBORINE (species of the genus *Epipactis*). XVI century (Gerard 1597, for White Helleborine, *E. latifolia*). From the Latin *helleborine*, from the Greek *helleborine*. The *helleborine* or *epipaktis* of the Greeks was taken to be *Veratrum album*, one of the plants which the Greeks also knew as *helleborus*. *E. latifolia* was called Helleborine because its leaves resemble the leaves of *Veratrum*.

HEMLOCK (*Conium maculatum*). Old English *hymlic*, one of the plant names ending in the suffix *-lic*, *-loc*, or *-ic*, *-oc*. There is no clue to the meaning of the name, which is only found in English.

HEMLOCK (Hemlock Spruce, Hemlock Fir, *Tsuga canadensis*, native of North America from Nova Scotia to Alaska, introduced 1736). Late XVIII century. From the resemblance of the branches to giant hemlock leaves.

HEMP (*Cannabis sativa*, native of temperate Asia, imported in the Middle Ages, grown by the early 16th century). Old English *hænep*, *henep*, a widely spread name in the Germanic languages for a plant cultivated in historic, but not prehistoric, Europe for its fibre. English *hemp* and Greek *kannabis* are cognate words.

HEMP AGRIMONY (*Eupatorium cannabinum*). XVIII century. A book name, combining old names. In the 15th century Hemp Agrimony was known as *canabria* (also Wild Hemp, Holy Rope), as if the Greek *kannabis agria*, 'wild hemp', mentioned by Dioscorides – 'this herbe hath lewys lyk to hemp.' It was also known to 16th century botanists as *cannabina aquatica*, the 'water hemp-like plant', and as a *eupatorium*, distinguished from *agrimonia* or *eupatorium* (our *Agrimonia eupatoria*, Agrimony) as *eupatorium adulterinum*, 'false eupatorium'.

HEMP-NETTLE (*Galeopsis tetrahit*). XVIII century, but XVI century as *Nettle Hemp* (Gerard 1597). Known to the 16th century botanists as *cannabis silvestris*, 'wild hemp'.

13. Green Hellebore (Brunfels, *Herbarum Vivae Eicones*, 1530)

HENBANE (*Hyoscyamus niger*). XIII century. *hen* + *bane*, 'hen killer', 'hen death'. In medieval Latin (7th century) *gallinaria herba* – possibly because as a casual *H. niger* occurs not infrequently on hen-scratched ground.

HENBIT (*Lamium amplexicaule*). XVI century, Gerard 1597. *hen* + *bite*. Cf. the Frisian name *hinnebyt*. Several low-growing plants went by the medieval Latin name *morsus gallinae*, 'hen's bite'.

HEPATICA (*Hepatica triloba*, native of the mountains of Europe, introduced in the 16th century). XVI century, Lyte 1578 – 'may be called in English *Hepatica*, Noble Agrimonie, or Three leafe Lyuervurte.' In medieval Latin (*herba*) *hepatica*, from the Latin adjective *hepaticus*, 'to do with the liver' (Greek *hepar*). 'The *Hepatica* . . . is a soueraigne medicine, against the heate and inflammation of the Lyver' (Lyte) – a use suggested by the liver-like appearance of the evergreen three-lobed leaves. See *Liverwort*.

HERB. XIII century. From the Old French *erbe*, from Latin *herba*. The meaning was at first 'grass', 'herbage', as distinguished from trees. So a plant for eating or taking medicinally.

HERB BENNET (*Geum urbanum*). XVI century, Lyte 1578 (earlier for, e.g., *Apium graveolens*). The apothecaries' name was *benedicta*, medieval Latin *herba benedicta*, 'blessed herb'. *G. urbanum* was regarded as powerful and blessed on account of its clove-scented root. As *herba benedicta*, and since it may be found in flower around the saint's feast-day, 21 March, it was associated also with St Benedict, St Bennet; and the root was used in the manufacture of Benedictine.

HERB CHRISTOPHER (*Actaea spicata*). XVI century. A translation of the medieval Latin name *herba Christofori*, 'herb of (St) Christopher'. Perhaps so named because *A. spicata* was held to be efficacious against the Plague, for which St Christopher was invoked, or because the flowers are borne in a spike above the leaves, like the infant Christ on St Christopher's shoulder.

HERB GERARD (*Aegopodium podagraria*, Goutweed, Ground Elder). XVI century, Lyte 1578, translating the *Gerardi herba* which was a Latinization of the Dutch name (for *A. podagraria*) *geraert*, German *Giersch*.

HERB OF GRACE (*Ruta graveolens*, Rue). XVI century, Turner 1548, *herbe grace*. Apparently a name consequent on the punning use of Rue, the bitter herb of ruth, of penitence, which was a state followed by the bestowal of the grace of God. So in Shakespeare's *King Richard II*, III, iv, 104,

> Heere did she drop a teare, heere in this place
> I'll set a Banke of Rew, sowre Herbe of Grace:
> Rue, ev'n for ruth, heere shortly shall be seene,
> In the remembrance of a Weeping Queene.

HERB PARIS (*Paris quadrifolia*). XVI century. Lyte 1578, translating the apothecaries' Latin *herba paris*, 'pair herb', 'herb of equality', from the numerical harmony of its parts, twice two leaves, twice four stamens, twice two outer and twice two inner segments to the perianth, twice two styles and twice two cells to the ovary.

HERB ROBERT (*Geranium robertianum*). XIII century. Translation of the medieval Latin *herba Roberti, herba (Sancti) Ruperti*. The origin of the name (in German *Ruprechtskraut*) may be in the redness of the plant, as if it had been known by some such name as *herba rubra*, this leading to association with St Rupert or Rudbert, of Salzburg, of the 8th century, who was invoked against bleeding wounds, ulcers and erysipelas, for which *G. robertianum* was given. Names of the plant also suggest the same kind of association with the house-goblin Knecht Ruprecht, the English Robin Goodfellow.

HIBISCUS (species of the genus *Hibiscus*; especially *H. syriacus*, in its many forms, from Eastern Asia, introduced c. 1596). XVIII century. English use of the Linnaean name, which was the Latin name for the Marsh Mallow (*Althaea officinalis*), from the Greek *hibiskos*.

HICKORY (trees of the genus *Carya*, from North America). XVII century. This very English-seeming tree name is a shortening of *pokahickory*, from Algonquian Indian *pawcohiccora*, milk from pounded hiccory nuts, a word recorded by the Virginia settler Captain John Smith (1580–1631).

HIP(s) (fruit originally of the Dog-rose). Old English *hēope*, fruit of the *hēopa*, 'the wild rose', cognate with the Old High German *hiufo*, and the Old Saxon *hiopo*, 'prickly bush' or 'bramble'.

HOG'S FENNEL (*Peucedanum officinale*, Sulphurwort). XVI century, 1585 (earlier Lyte 1578 *sowefenill*), translating the German *Saufenchel*. See also *Fennel*.

HOGWEED (*Heracleum sphondylium*, Cow-parsnip). XVIII century. Much collected to feed to pigs.

HOLLY (*Ilex aquifolium*). XII century *holi*, Old English *holegn*. A Germanic name, cf. German *Hülse, Hulst*; and Old High German *hulis*.

HOLLYHOCK (*Althaea rosea*, from China, introduced in the 16th century). XIII century, for the European and English healing herb, Marsh Mallow, *Althaea officinalis. holy + hock* (Old English *hocc*), a mallow – holy evidently as a blessed or healing herb, as in Holy Thistle (*Cnicus benedictus*), the *benedicta, carduus sanctus, carduus benedictus*, of early botanists. When the Chinese *Althaea rosea* reached England in the 16th century, Turner 1548 gave it the old name hollyhock (*Holyoke*); *Althaea officinalis* he called *marrishe Mallowe*. Lyte 1578 calls the Chinese plant *Holyhocke, garden Mallow*, and *beyondsea rose*.

HOLY ROSE (*Rosa × richardii*, Abyssinian Rose, probably from Asia Minor, introduced from Ethiopia c. 1902). XX century. A rose which was discovered in the forecourts of churches in the province of Tigre in Ethiopia.

HONESTY (*Lunaria annua*, native of S.E. Europe, introduced in the 16th century). XVI century, Gerard 1597: 'among our women it is called Honestie.' The name was also given, 17th century, to Old Man's Beard (*Clematis vitalba*).

HONEY FUNGUS (*Armillaria mellea*). XIX century. Translating the *mellea* of its specific name, Latin *melleus*, adjective of *mel*, 'honey'. The fungus is honey-yellow.

HONEYSUCKLE (*Lonicera periclymenum*, Woodbine). XVI century, as *L. periclymenum*. Earlier, Old English *honigsūge* (*sūgan*, to suck), 13th century *hunisuccle*, for Red Clover (*Trifolium pratense*). From the pleasurable sucking of the honey from the corolla tube.

HOP (*Humulus lupulus*). XV century, from the Dutch *hoppe*, from Old High German *hopfo*. The basic meaning of this Germanic name of a plant with twisting exploring tendrils may be 'that which gropes or fumbles around'.

HOREHOUND (*Marrubium vulgare*). *hār hūne* in Old English, *hār*, 'hoar', 'grey', 'pubescent', + the plant name *hūne*.

HORNBEAM (*Carpinus betulus*). Old English *horn-bēam*, horn (owing to the hardness of the timber) + *bēam*, tree. The name is not recorded until the XVI century, though it occurs in place-names two centuries earlier.

HORSE-. In plant names this frequently indicates some coarse differentiating quality.

HORSE-CHESTNUT (*Aesculus hippocastanum*, native of S.E. Europe introduced in the late 16th century). XVI century, Gerard 1597. Translation of the botanists' Latin *castanea equina*, a coarse chestnut, but Matthiolus (1501–1577) in the first description of *A. hippocastanum* wrote that the Turks call the fruits horse-chestnuts because they helped horses with a difficulty in breathing.

HORSE-MINT (*Mentha longifolia*, and other wild species of mint). XIII century. See *Horse-* above.

HORSE-MUSHROOM (*Agaricus arvensis*). Not on record before the XIX century. On account of the large size and the yellow tinted cap, as if less good than the Field-mushroom. See *Horse-* above.

HORSE-RADISH (*Armoracia rusticana,* native of Western and S.E. Asia). xvi century, Gerard 1597. The name is a mingled translation of the early botanists' Latin *raphanus rusticanus, raphanus sylvestris,* 'country' or 'wild radish', and the German *Meerettich,* as if *Mähre,* 'mare', + *Rettich,* 'radish'; whereas the German name in fact means sea-radish (*Meer*), in the sense of a radish or root from foreign parts, a 'beyondsea radish' (cf. Lyte's 'beyondsea rose' for the foreign Hollyhock). *Horse-radish* was also well suited in English practice to describe a large, coarse, strong and strong-scented plant. See *Horse-* above.

HORSETAIL (*Equisetum arvense,* etc.). xvi century, Turner 1538, translating the medieval Latin *cauda equina,* descriptive of the stems and branches.

HOTTENTOT FIG (*Carpobrotus edulis,* native of South Africa, introduced 1690). xviii century. Fruit eaten by the black people of South Africa.

HOUND'S-TONGUE (*Cynoglossum officinale*). Old English *hundes tunge,* translating the medieval Latin name *lingua canis,* translating in turn the Greek *kunoglosson,* with which *C. officinale* was equated. From the shape of the leaves.

HOUSELEEK (*Sempervivum tectorum,* anciently introduced as a medicinal plant and roof-protection against lightning. Unknown as a wild plant). Old English *hūslēac.* There are similar names in other Germanic languages.

HUCKLEBERRY (species of the genus *Gaylussacia,* natives of North America; and the fruits). xviii century, usually taken as a variant of Hurtleberry (the related Whortleberry or Bilberry, *Vaccinium myrtillus*). The derivation of *Hurt, Hurtle(berry),* is unknown. See *Bilberry, Whortleberry.*

HURTLEBERRY. See *Huckleberry,* above, and *Whortleberry.*

14. Hop (Fuchs, *De Historia Stirpium*, 1542)

HYACINTH (*Hyacinthus orientalis*, native of Greece and Asia Minor, introduced in Elizabethan times). XVI century, Lyte 1578, 'the Oriental Hyacinthes'. From the French *hyacinthe*, from the Latin *hyacinthus*, from the Greek *huakinthos*, which appears to have been, not *H. orientalis*, but the little spring-flowering *Scilla bifolia* of S.E. Europe (though the name was also given to the Larkspur, *Consolida ambigua*), called after the vegetation god Huakinthos, whose name is pre-Hellenic. The Greek myth is that the hyacinth sprang from the blood of Huakinthos when he was accidentally killed by the discus of Apollo. The proper hyacinthine colour is blue – the colour of the precious stone *huakinthos* (probably the aquamarine), of the flowers of *Scilla bifolia* and *Consolida ambigua* (and also of the unimproved *Hyacinthus orientalis*). So the name was transferred to the Wild Hyacinth, the Bluebell, of Atlantic Europe (*Endymion nonscriptus*), which Gerard in the 16th century called *Hyacinthus Anglicus* and *English Jacint* (= hyacinth). See also *Bluebell*.

HYDRANGEA (species of the genus *Hydrangea*, especially *H. macrophylla*, native of China and Japan, introduced in 1740). XVIII century. English use of the Linnaean name (1737), Latinized from Greek *hudor*, 'water', and *aggeion*, 'jar', 'vase', from the shape of the capsule.

HYSSOP (*Hyssopus officinalis*, native of the Mediterranean region, introduced in the Middle Ages). Old English *ysope*, from Latin *hyssopus*, from Greek *hussopos*, from Hebrew *ēsōb*. But the hyssop of the Romans, the Greeks and the Bible appears to have been Marjoram (*Origanum vulgare*).

I

ICE-PLANT (*Cryophytum crystallinum*, native of the Mediterranean region, introduced 1775). XVIII century. From its icy, crystalline appearance.

ICELAND MOSS (the medicinal lichen *Cetraria islandica*). 1805, translating *muscus islandicus*, by which *C. islandica* was known to early botanists and apothecaries.

ILEX (*Quercus ilex*, Holm-oak, Evergreen Oak, native of the Mediterranean region, introduced in the 16th century). XVI century. The Latin name, as in Virgil's *Eclogues*.

IMMORTELLE (species of various mainly Composite genera, whose dried flowers retain shape and colour). XIX century. French *Immortelle*, shortened from *Fleur immortelle*, 'everlasting flower', in French mainly *Helichrysum orientale*, from South Africa. See *Everlasting*.

INDIA RUBBER. See *Rubber*.

INDIAN CORN (*Zea mays*, Maize, of American origin, introduced in the 16th century). XVII century. First on record in the second edition, 1622, of Captain John Smith's *New Englands Trials . . . with the benefit of that Countrey by sea and land.*

INDIGO (blue dye from the tropical plant *Indigofera tinctoria*). XVI century, *indico*, from Spanish *indico*, from Latin *indicum*, from Greek *indikon* (*pharmakon*), 'Indian dye'.

INKY CAP (the mushroom *Coprinus comatus*). XX century. Because it deliquesces into a black 'ink'.

IPECACUANHA (*Cephaelis ipecacuanha*, from Brazil; and its medicinal rhizome introduced to Europe in 1672, in the

109

treatment of dysentery). XVII century. Portuguese *ipecacuanha*, adapted from Tupi *ipekaaguéne*, 'low plant which makes sick'.

IRIS (species of the genus *Iris*, especially the wild *I. pseudacorus*). XVI century, Lyte 1578, 'the Dwarffe Ireos, the stincking Iris, and the yellow Iris'. From the Latin *iris*, from Greek *iris*, 'the rainbow'; describing the diversified colouring of its flowers (Pliny).

IRISH MOSS (the seaweed *Chondrus crispus*, Carragheen). 1845. On the analogy of Iceland Moss, for which *Chondrus crispus* was promoted as a medicinal equivalent.

IVY (*Hedera helix*). Old English *ifig*, with related names in other Germanic languages.

J

JACARANDA (trees of the genus *Jacaranda*, natives of Brazil). XVIII century. Portuguese *jacarandá*, from the Tupi-Guarani name.

JACK-BY-THE-HEDGE (*Alliaria petiolata*, Sauce-alone, Garlic Mustard). XVI century, Turner 1538: 'sauce alone, aut ut alii vocant Jak of the hedge'. Lyte 1578, *Jacke by the Hedge. A. petiolata* is a roadside, hedge-colonizing plant.

JACK-GO-TO-BED-AT-NOON (*Tragopogon pratensis*). XVI century, Lyte 1578, *Go to bedde at noone.* The flowers shut at midday.

JACOB'S LADDER (*Polemonium coeruleum*). XVIII century. From the ladder of Jacob in *Genesis* xxviii, 12: 'And he dreamed, and behold a ladder set up on the earth, and the top of it reached to heaven,' on account of the ladder-like arrangement of the pinnate leaves, surmounted by blue flowers.

JALAP (loosening drug from tubers of the Morning Glory type plant *Exogonium purga*, native of Mexico). XVII century. From the French *jalap*, from Spanish (*purga de*) *Jalapa*, 'purge from Jalapa', mountain city in Mexico, in a district where the plant grows.

JAPONICA (*Cydonia speciosa*, native of China, introduced 1815. Japanese·Quince). XIX century; but first used of the Camelia (*Camellia japonica*), introduced 1739. The earliest mention of Japonica in the *Oxford English Dictionary* is from a letter Keats wrote to his sister in March 1819, in which he imagines a globe of goldfish, shaded by myrtles 'and Japonicas', i.e. camelias, in a window looking out on the Lake of Geneva.

JARGONELLE (variety of pear). XVII century, Evelyn 1693. French *jargonelle*, diminutive of *jargon*, the precious stone zircon.

JASMINE (*Jasminum officinale*, native of Persia, introduced in the 16th century). XVI century, Turner 1562, *Jessamin*; Lyte 1578 *Jasmine*. 'It is called amongst the herboristes of England, Fraunce, and Germanie *Iasminum*, and *Ieseminum*' – from the Arabic *yāsamīn*, from the Persian *yāsemīn*.

JENNETTING, JOANETING, JUNETING (variety of apple). XVII century. An apple name familiar since Elizabethan times. From the *pomme de Jeannet* of the Normandy apple-growers (*Jeannet*, pet-form of the name *Jean*, + the English collective suffix -*ing*).

JERUSALEM ARTICHOKE (*Helianthus tuberosus*, native of U.S.A. and northern South America, introduced into Europe early in the 17th century, into England 1617). The Italian botanist Fabio Colonna wrote in 1616 of *H. tuberosus* growing in Rome in the Farnese garden. The new vegetable was recognized as a sunflower with tubers which tasted something like the edible parts of the Globe Artichoke (*Cynara scolymus*). The Italian *girasole*, sunflower (*girare*, 'to turn round', + *sole*, 'sun'), was garbled in English as 'Jerusalem' – Jerusalem Artichoke, first recorded (*Oxford English Dictionary*) in 1620 as *Artichocks of Ierusalem*, which at least suggested the exotic origin of *H. tuberosus*. Colonna had called the plant *Aster peruvianus tuberosus*, as if from Peru. Other early names were *Chrysanthemum ex Canada*, as if from Canada (in Italian *H. tuberosus* is called *girasole di Canada*), and *helianthemum indicum tuberosum*, as if from the West Indies; also *flos solis Farnesianus*, Farnesian sunflower. See also *Artichoke*.

JERUSALEM COWSLIP (*Pulmonaria officinalis*, Lungwort, native of Central and Northern Europe and the Caucasus, introduced probably in the 16th century). XVI century, Lyte 1578, 'We call it in English Sage of Ierusalem, and Cowslip of Ierusalem' – Jerusalem, no doubt, for an outlandish plant. Lyte also says that the plant bears at the top flowers 'growing togither in a bunch like Cowslip floures'.

JERUSALEM SAGE (*Phlomis fruticosa*, native of the Mediter-ranean region. Introduced late in the 16th century). XVIII century. *Jerusalem-Sage*, or *Sage-tree*. 'Sage of Jerusalem' was used in 16th century (Lyte 1578) for Lungwort, *Pulmonaria officinalis* (see above, Jerusalem Cowslip).

JEW'S EAR (the fungus *Auricularia auricula*). XVI century. Translation of the botanists' and apothecaries' Latin name *auricula Judae*, 'ear of Judas', as if *auricula judaea*, 'Jewish ear'. This ear-like fungus grows on the elder, on which Judas was supposed to have hanged himself.

JIMSON-WEED (*Datura stramonium*, Thorn-apple). XVII century. The weed from Jamestown, Virginia. (The American name for *D. stramonium*.)

JOB'S TEARS (the grass *Coix lacryma-jobi*, native of tropical Asia, introduced in the late 16th century). XVI century, Gerard 1597, 'Iobs Teares or Iobs Drops', translating the botanists' Latin *lachryma Jobi*. From the shining seeds, and *Job* xvi, 16 and 20, 'My face is foul with weeping . . . Mine eye poureth out tears unto God.'

JONATHAN (American variety of apple, raised in New York State). XIX century. Probably an assimilation to the name Jonathan of the old apple name Jennetting, Joaneting, Juneting. See *Jennetting*.

JONQUIL (*Narcissus jonquilla*, native of Southern Europe and Algeria, introduced in the late 16th century). XVII century. From French *jonquille*, which is from the Spanish *jonquillo*, diminutive of *junco*, 'rush', in reference to the rush-like leaves.

JUDAS-TREE (*Cercis siliquastrum*, native of Southern Europe and Western Asia, introduced in the late 16th century). XVII century. Translation of the early botanists' Latin *arbor Judae*, from the legend that Judas hanged himself on a Judas-tree, which in penitence for his betrayal of Christ and sorrow for the Passion of Christ produces its flowers like blood, before the leaves.

JUJUBE (soft edible fruit of *Zisyphus jujuba*, native of S.E. Europe to China; and the tree). XIV century. From Old French *jujube*, from medieval Latin *jujuba*, from Latin *zizyphum*, from Greek *zizuphon*. Pliny writes of jujube-trees as recently introduced into Italy from Africa.

JUNIPER (*Juniperus communis*). XIV century. From the Latin *juniperus*, which has been referred to the same Indoeuropean base as the Latin *juncus*, a rush, with the common meaning of something used for binding – the tough stems of rushes, and the tough branches of the juniper.

K

KALE, northern form of *cole* (q.v.).

KELP (seaweeds including the Wracks, *Fucaceae*, and Oarweeds, *Laminariaceae*, which strew beaches after a storm). xiv century, *culp*; of unknown origin.

KERMES OAK (*Quercus coccifera*, native of the Mediterranean region and Western Asia, introduced 1683). xvi century. Oak on which the kermes insect is found, the insect giving the red dye, from French *alkermès*, from Arabic and Persian *qirmiz* (source of the word *crimson*), from Sanskrit *krmi-dza*, 'made by a worm'.

KEX (hollow, dry stalks of umbelliferous hedge plants such as Hogweed, *Heracleum sphondylium*). xiv century. ? through British from the Latin *cicuta*, hemlock (cf. Welsh *cegid, cegr*).

KIDNEY BEAN (*Phaseolus vulgaris*, French Bean, native of America, introduced in the 16th century). xvi century, but first for *Phaseolus coccineus*, the Scarlet Runner: Turner 1548, 'It may be called in english Kydney beane, because the seede is lyke a kydney, or arber beanes, because they serve to cover an arber for the tyme of Summer.'

KIDNEY VETCH (*Anthyllis vulneraria*). xviii century. The plant was used against kidney troubles. Lyte 1578, 'If one drinke halfe an ounce of the first *Anthyllis*, it shall prevayle much against the hoate pisse, the strangury or difficultie to make water, and against the payne of the Reynes.'

KINGCUP (*Caltha palustris*, Marsh Marigold). xvi century. A name, first recorded by Turner 1538 for one of the Buttercups (*Ranunculus bulbosus*), 'Kyngcuppe, aut a Golland', applied widely in the country to pasture buttercups and to *C. palustris*.

KNAPWEED (*Centaurea nigra*, Hardheads). xv century, *knopwed*; *knop*, 'head', + *weed*, in reference to the hard flower-heads. *C. nigra* has similar *knop* names in German, Swedish and Dutch.

KNAWEL (*Scleranthus annua*). xvi century, Lyte 1578, 'The thirde kinde [of knot grass] is called in base Almaigne knawel.' (German *Knäuel*, ball of thread, tangle.)

KNOTGRASS (*Polygonum aviculare*). xvi century, Turner 1538, 'knotgyrs'. Descriptive of the way the stem seems to advance from knot to knot.

KOHLRABI (*Brassica caulorapa*, Stem Turnip). xix century, German *Kohlrabi*, from the Italian plural *cavoli rape*, 'cabbage turnips'. In German the Swede (*Brassica napus* var. *napobrassica*) was also a *Kohlrabi unter der Erde*, 'under the earth', as distinguished from the Stem Turnip, *Kohlrabi über der Erde*, 'above the earth'.

KUMQUAT (species of *Fortunella*, natives of S.E. China; and the fruit). xvii century, *cam-quit*. Cantonese *kam kwat*, 'gold orange' (Middle Chinese *kyĕm kywit*).

L

LABLAB (*Dolichos lablab*, native of Asiatic and African tropics). XIX century. Turkish *lablāb*, Arabic *lubia*.

LABURNUM (*Laburnum anagyroides*, native of Central and Southern Europe, introduced in the 16th century). XVI century, Lyte 1578. From the Latin *laburnum* (Pliny, who says, 'A tree from the Alps, not widely known, with hard white wood, and flowers eighteen inches long which bees will not touch'). Possibly a loan-word from Etruscan.

LAD'S LOVE (*Artemisia abrotanum*, native of Southern Europe, introduced in the early Middle Ages). XIX century. The name is explained by the ancient reputation of *A. abrotanum* as an aphrodisiac, to which 16th century herbalists refer (e.g. Fuchs, Brunfels). It was also held to produce abortions, bring on the curse, and in general to open the private parts of women (Gösta Brodin, *Agnus Castus*, 1950, p. 209, quoting a Latin herbal written in England in the late 14th century). Lyte 1578 adds to his account of *A. abrotanum*, 'Plinie writeth that if it be layde under the bedde, pillow or bolster, it provoketh carnall copulation, and resisteth all enchantments, which may let or hinder such businesse and the inticements to the same.' See also *Southernwood*.

LADIES-, LADY-, LADY'S-. In plant names usually a shortening of *Our Lady's*, for a plant of the Virgin Mary, often one of medicinal reputation. Such names seem mostly to have been translated from German equivalents in herbals of the 16th century, and they no doubt originated in monasteries in the late Middle Ages. None are recorded in English (or German) before the 16th century, though *Costmary, Seynt Mary maythe* for Oxeye Daisy (*Chrysanthemum leucanthemum*), *Seynt marie seal* for Solomon's

Seal (*Polygonatum multiflorum*), and *Seynt Marie seed* for the seed of *Sonchus oleraceus* (Sow-Thistle) or *Silybum marianum* (Lady's Thistle) occur in the 15th century; and *Marygold* (q.v.) in the 14th century.

LADY FERN (*Athyrium filix-femina*). XIX century. The female fern of the 16th century botanists, in contrast to their Male-fern (*Dryopteris filix-mas*) was the Bracken (*Pteridium aquilinum*). Linnaeus made *A. filix-femina* the female fern, calling it *Polypodium filix-femina*; and *filix-femina* was Englished in the 19th century as Lady-fern, in deference, not to the Virgin Mary, but to the elegance of the fern (cf. Thomas Moore in *A Popular History of British Ferns*, 1851: 'The Lady Fern, on account of the exquisite grace of its habit of growth, the elegance of its form, and the delicacy of its hue, claims precedence over every other British species'). Early 19th century writers occasionally use Lady Fern for Bracken. See also *Male Fern*.

LADY-SMOCK, LADY'S SMOCK (*Cardamine pratensis*, Cuckoo Flower). XVI century, Gerard 1597, 'Ladie smockes', from which Shakespeare may have taken his 'Ladi-smockes all silver white', in Ver's song at the end of *Love's Labour's Lost*, in the Quarto, 1598, 'Newly corrected and augmented'. Probably this is not one of the names with (*Our*) *Lady* (see above, under *Lady*). The early spellings are Lady-smock (not Lady's or Ladies Smock), which may be an assimilation to the form of *Our Lady* plant names or some unrecorded survival of the Old English *lustmoce* (*C. pratensis*), the meaning of which seems as little edifying as the meaning of cowslip (q.v.) – if the first element is Old English *lust*, sexual desire, and the second related to *moc*, 'muck', (as in Old English *hlōs moc*, 'pigsty muck'). The name would then refer to the semen-like spume or cuckoo-spit surrounding the larva of the froghopper (*Philaenus spumarius*) so generally found on plants of Lady-smock.

LADY'S BEDSTRAW (*Galium verum*). XVI century. Translation of the German *Unser Frawen Betstro*. First attested, but for Wild Thyme (*Thymus serpyllum*), in 1527, in a translation from the

German. In the German herbals of the 16th century, it is the regular name for *G. verum*, and, in their wake, Lyte 1578 has 'our Ladies bedstraw'. The name comes from the legend that the bedstraw on which the Virgin lay was a mixture of bracken and *G. verum*: the Bracken did not acknowledge the child when it was born, and lost its flowers. The *G. verum* welcomed the child, and blossomed then and there, finding that in Christ's honour its flowers had changed from white to yellow.

LADY'S CUSHION (*Armeria maritima*). XVI century. Lyte 1578, 'some call it in Englishe our Ladies quishion.'

LADY'S FINGERS (*1. Anthyllis vulneraria*, Kidney-vetch; *2.* pods of *Hibiscus esculentus*, Okra, Gumbo, native of Africa). In the first sense, XVII century (Ray 1670). German names for *A. vulneraria* (which was celebrated as a wound-herb) include *Unser Frauenschüh*, *Herrgottsschüle*, 'Our Lady's Shoe', 'God's Little Shoe' (i.e. the shoe of the infant Christ). In the second sense, XX century, from the shape and softness of the pods.

LADY'S GLOVE (*Digitalis purpurea*). XVII century. A translation of the botanists' Latin *Chirotheca Sanctae Mariae*, or French *gant de Notre-Dame*, which was also given to *Campanula trachelium*, Throatwort, and *C. medium*, the Canterbury Bell. Earlier, 'our ladies gloves' (Elyot 1538), in uncertain application.

LADY'S MANTLE (*Alchemilla vulgaris*). XVI century, Turner 1548, 'our Ladies Mantel', in reference to the mantle-shaped leaves, translating the *Unser Frawen mantel* of the German 16th century herbalists.

LADY'S SEAL (*1. Polygonatum multiflorum*, Solomon's Seal; *2. Tamus communis*, Black Bryony). XVI century. In the first sense, *Grete Herball* 1526, translating the medieval Latin name *Sigillum sanctae Mariae*. Earlier *seynt marie seal* (15th century). In the second sense Lyte 1578 (cf. French *Sceau de la Vierge, Sceau de Notre-Dame*).

LADY'S SLIPPER (*Cypripedium calceolus*). XVI century, Gerard 1597, 'Our Ladies Shooe or Slipper'. From the shape of the flowers, translating the botanists' Latin *calceolus Mariae* or German *Unser Frawenschuh*.

LADY'S THISTLE (*Silybum marianum*, Milk Thistle). XVI century, Elyot 1552, '*Our ladies Thistle*'. Probably a translation of the German *Frawen Distel*. See *Milk Thistle*.

LADY'S TRESSES (*Spiranthes spiralis* and *S. aestivalis*). XVI century. Turner 1548, 'a certeyne ryghte kynde of the same groweth besyde Syon, it bryngeth furth whyte floures in the ende of harueste, and it is called Lady traces.' *trace*, 'a braid of hair', or 'tress'. The flowers are arranged in a braid-like spirally twisted row.

LAGER (light beer, especially from Bavaria, Holland, Denmark). XIX century. From German *Lagerbier*, 'store beer', i.e. beer kept in store before drinking.

LAMB'S LETTUCE (*Valerianella locusta*, Corn Salad). XVI century. Gerard 1597, translating the *Lactuca agnina* of the 16th century herbalists.

LANDLADY'S WIG (the seaweed *Desmarestia aculeata*). XIX century. In reference to the brownish tangle of fronds thrown up on beaches.

LARCH (*Larix decidua*, native of the Alps, introduced early in the 17th century). XVI century. Turner 1548, translating the German *Lärche*: 'Larix or larex groweth in the highest toppes of the Alpes higher than the firres do, the duch men cal Laricem ein larchen baum . . . It may be called in englishe a Larch tree.' *Lärche*, from Old High German *larihha, from *laricem*, accusative of Latin *larix*, probably a loan-word from a language spoken in the Alpine region.

LARKSPUR (*Delphineum ajacis*, native of Southern Europe, introduced in mid-16th century). XVI century, Lyte 1578. From the long spur of the posterior sepal. An Englishing of French *pied d'alouette* or *éperon d'alouette*; or German *Lerchenklaue*.

LAUREL (now generally *Prunus laurocerasus*, Cherry-Laurel, native of S.E. Europe and S.W. Asia, introduced early in the 17th century; earlier *Laurus nobilis*, Bay-tree, native of the Mediterranean region, introduced in the 16th century). In the earlier sense XIII century, *lorer* (*lorel* XIV century), from Old French *lorier*, earlier *lor*, from Latin *laurus*, which was probably a loan-word from a Mediterranean language. In the later sense, XVIII century, short for Cherry Laurel.

LAURUSTINUS (*Viburnum tinus*, native of the Mediterranean region, introduced at the end of the 16th century). XVII century. English use of the botanists' Latin *Laurus tinus* (Parkinson 1629), *Laurus* as a Wild Bay (Parkinson's synonym for *V. tinus*) + *tinus*, which Pliny mentions as a kind of laurel.

LAVENDER (*Lavandula officinalis*, native of the Mediterranean region, introduced in the 16th century). XIII century, from Anglo-Norman *lavendre*, from the medieval Latin *lavendula*, from Italian *lavandula*, a diminutive of Italian *lavanda*, 'that which serves for washing, laving' (Latin *lavare*, to wash), i.e. the plant used to perfume water for washing.

LAVENDER COTTON (*Santolina chamaecyparissus*, native of the Mediterranean region, introduced in the 16th century). XVI century. From the cottony-white look and lavender-like growth of the plant, which was grown in gardens for such domestic purposes as ridding children of worms.

LAVER (the edible seaweed *Porphyra umbilicalis*). XVII century. English use of the Latin *laver*, a fresh water plant mentioned without definition or description by Pliny.

LAWYER'S WIG (the mushroom *Coprinus comatus*, Shaggy Cap, Inky Cap). XX century. From the rounded cap and shaggy white scales.

LEEK (*Allium porrum*, of uncertain origin). Old English *lēac*, leek, garlic. A Germanic name with cognates in German, Dutch, Swedish, etc., from an Indoeuropean base meaning to twist or bend. See *Garlic*.

LEMON (*Citrus limonia*, native of Eastern Asia; the fruit and the tree). XIV century (the fruit). From the Old French *limon*, from the Italian *limone*, from Turkish *limon*, from Persian *līmūn*.

LEMONADE. XVII century. From the French *limonade* (17th century). The earliest quotation in the *Oxford English Dictionary* is dated 1663.

LENTIL (*Lens esculenta*, native of Southern Europe; the seed and the plant). XIII century. From Old French *lentille*, from Latin *lenticula*, diminutive of *lens*, accusative *lentem*, with the same meaning.

LENT LILY (*Narcissus pseudonarcissus*, Daffodil). XIX century. Cf. Dutch *paaschlelje*, Danish *påskelilie*, German *Paschlelech*, 'Pasch lily', 'Easter lily'.

LEOPARD'S BANE (*Doronicum pardalianches*, native of Western Europe, introduced at the end of the 16th century). XVI century. An equivalent for the Greek *pardaliagkhes*, 'that which strangles the leopard', name of a plant described by Dioscorides, the synonym for which was *akoniton*. Turner 1548 coined Leopards Bane ('Libardbayne'), believing that the Dioscoridean plant was Herb Paris (*Paris quadrifolia*). Other herbalists took the extremely poisonous Dioscoridean plant to be the innocent and pretty *D. pardalianches* of Alpine woods and European gardens. Lyte 1578 used Leopards Bane for *D. pardalianches*, and the name has persisted.

LETTUCE (*Lactuca sativa*, ? from Siberia or S.W. Asia). XIII century, *letus*, from the plural, *laitues*, of Old French *laitue*, from Latin *lactuca*, *lac* (genitive *lactis*), 'milk', + noun suffix, 'milky plant'.

LIANA (any tropical twining plant, or its vine). XVIII century, *lianne*; 19th century, *liana*, an alteration as if Spanish and so fitting the word to its tropical New World association. From the West Indian French *liane*, a tie around a sheaf or faggot, and so any plant stem which can be used in that way, from French *lien* in the same sense (*lien de fagot, lien de gerbe*), from *lier*, 'to bind', from Latin *ligare*.

LIBERTY CAP (the fungus *Psilocybe semilanceata*). XIX century. From the resemblance of the pointed caps to the conical Phrygean bonnet or cap of liberty, given to Roman slaves when they were emancipated.

LICHEN (plant of the group *Lichenes*). XVII century. English use of the Latin *lichen*, from Greek *leikhen*, 'lichen', 'liverwort'.

LIGNUM VITAE (wood of the tree *Guaiacum officinale*, native of the West Indies and northern South America). XVI century. From the apothecaries' Latin, meaning 'wood of life', *Lignum vitae* having been famous, from the 16th century, for its use in treating syphilis (the less euphemistic 16th century name was Pockwood).

LILAC (*Syringa vulgaris*, native of S.E. Europe, introduced at the end of the 16th century). XVII century. From French or Spanish *lilac*, from Arabic *lilak*, a loan-word from Persian. Persian *lilak*, 'blueish plant or flower', is a form of the word *nilak*, 'blueish', from *nil*, 'indigo'. The earlier 17th century name was Pipe-tree, the Pipe-trees including both the lilac and the Syringa (*Philadelphus coronarius*). See *Syringa*.

LILY (species of the genus *Lilium* or the family *Liliaceae*). Old English *lilie*, from the Latin *lilium*, related to Greek *leirion*. The conjectured source of both the Greek and Latin words is the Egyptian *hrr-t*, pronounced in Coptic *hlēli*, *hrēri*.

LILY-OF-THE-VALLEY (*Convallaria majalis*). XVI century. An Englishing of the medieval Latin *lilium convallium* in the Vulgate translation of the *Song of Songs* ('I am the rose of Sharon, and the lily of the valleys', in the Authorized Version), the *lilium convallium* having been equated with *C. majalis* by 16th century apothecaries (Turner 1548: 'The Poticaries in Germany do name it lilium convallium, it maye be called in englishe May Lilies').

LIMA BEAN (*Phaseolus limensis*, native of tropical America, the bean and the plant, introduced in the 18th century). XIX century. Associated with Lima, the Peruvian capital.

LIME, LIME TREE (*Tilia* × *vulgaris*, introduced). XVII century. An alteration of *line*, which was a 16th century variant of *lind*, from Old English *lind*, which, with other Germanic cognates, may descend from an Indoeuropean base meaning pliable (the reference being to the fibres of the inner bark used for weaving into ropes, halters, etc. The bark 'next to the timber . . . will easily be wrested, turned, and twisted every way', Lyte 1578).

LIME (fruit of *Citrus aurantifolia*, native of S.E. Asia and India). XVII century. From the French *lime*, from modern Provençal *limo*, from Arabic *līmah*. So *lime-juice* (18th century).

LINDEN, LINDEN TREE (*Tilia* × *vulgaris*, Lime Tree, introduced). XVI century. From the German *Linden(baum)*.

LING (*Calluna vulgaris*, Heather). XIV century. From the Old Norse *lyng*. The common name in the English areas of Danish or Norwegian settlement.

LIQUIDAMBAR (*Liquidambar styraciflua*, Sweet Gum, native of North America and Mexico, introduced 1683). XVI century, for the gum or resin of *L. styraciflua*; XVII century, *Liquid-amber* or *Liquid Amber*, for the tree. For the fragrant gum the Latin *liquidambar* (*liquidus*, 'liquid', + medieval Latin *ambra*, 'amber') was coined by the 16th century Spanish physician Francisco Hernandez. The gum was sold in apothecaries' shops as *Balsamum liquidambrae*.

LIQUORICE (*Glycyrrhiza glabra*, native of Central Europe and the Mediterranean region, introduced in the 16th century). XIII century, for the rhizome or the preparation from the rhizome; XVI century, for the plant. From Anglo-Norman *lycorys*, from Old French *licorice*, from Late Latin *liquiritia*, rendering Greek *glukurrhiza* (*glukis* 'sweet', + *rhiza*, 'root').

LITCHI, LICHEE (fruit of *Litchi chinensis*, native of China). XVI century, *lechia*; XVII century, *lichee*. From the Pekin Mandarin *li-chih*.

LIVELONG (*Sedum telephium*, Orpine, Midsummer Men). XVI century, Lyte 1578, who calls it 'Liblong, or Live long': 'The people of the countrey delight much to set it in pots and shells on Midsomer Even, or upon timber slattes or trenchers dawbed with Clay, and so to set, or hang it up in their houses, where as it remayneth greene a long season and groweth if it be sometimes over sprinckled with water.' See *Midsummer Men*.

LIVERWORT (species of the class *Hepaticae*, especially *Marchantia polymorpha*). XII century, for a plant used against liver complaints; XV century, for *M. polymorpha*, from the lobed liverlike thallus which suggested such use. See *Hepatica*.

LIZARD ORCHID (*Himantoglossum hircinum*). XVIII century. From the resemblance of the long central lobe of the flower to a lizard. Also Lizard Flower (Miller 1731). Lyte 1578 had remarked on the many flowers 'much like to a Lezarde, bycause of the twisted or wrythen tayles, and speckled heades', but he gave no English name.

LOBELIA (species of the genus *Lobelia*, especially the small edging plant *Lobelia erinus*, native of South Africa, introduced 1752). XVIII century. English use of the botanists' Latin *Lobelia* (1703) given in honour of the great Flemish botanist Lobelius – Mathias de l'Obel (1538–1616), Botanist to James I of England.

LOBSTER HORNS (the seaweed *Polysiphonia elongata*). XIX century. From the likeness of the branchlets to lobster's antennae.

LOCO-WEED (*Oxytropis lambertii*, native of the United States). XIX century, 'mad-weed', from Spanish *loco* 'mad'. The plant causes a form of madness in horses.

LOCUST TREE (*Robinia pseudoacacia*, native of the Eastern and Central United States, introduced c. 1836–1840). XVII century, *Locus, Locus tree*. ? of Indian origin, a name changed to *Locust*,

Loddon Lily

Locust Tree under the influence of the Locust Tree (= *Ceratonia siliqua*) of the Old World.

LODDON LILY (*Leucojum aestivum*, Summer Snowflake). XIX century. Because of the abundance of *L. aestivum* along the river Loddon in Berkshire.

LODGEPOLE PINE (*Pinus contorta* var. *latifolia*, from the Rocky Mountains, introduced about 1853 and much planted in damp western areas of Britain). XX century. The species is straight-growing and was used by North American Indians to hold up their skin-covered lodges.

LOGANBERRY (*Rubus loganobaccus*, of Californian origin). XX century (the earliest quotation in the *Oxford English Dictionary* is dated 1900). From the American judge James Harvey Logan (1841–1928), who raised the loganberry by hybridization in 1881.

LOGWOOD (heartwood of *Haematoxylum campechianum*, native of the West Indies and Central America; and the tree). XVI century, for the heartwood logs imported for the dyers of cloth; XVII century, for the tree.

LOMBARDY POPLAR (*Populus italica*, native of Russian Central Asia and Afghanistan, introduced in 1758 from Italy). XVIII century, when *P. italica* was also known as the *Po-poplar*, from its frequency along the great river dividing the Lombardy plain.

LONDON PLANE (*Platanus* × *acerifolia*, hybrid between *P. orientalis* from S.E. Europe and Western Asia, introduced early in the 16th century, and *P. occidentalis* from North America, introduced from Virginia in 1636). XIX century. Planted for more than two centuries in London streets and squares.

LONDON PRIDE (now the garden hybrid *Saxifraga spathularis* × *umbrosa*, of unknown origin; earlier a form of the Sweet William, *Dianthus barbatus*). In the earlier sense, XVII century (Parkinson 1629); in the present sense, late XVII century.

LONDON ROCKET (*Sisymbrium irio*, native of the Mediterranean region). XIX century. A botanists' book name given to this casual because it had appeared abundantly on derelict areas after the Great Fire of London in 1666.

LONG PURPLES (*1. Orchis mascula*, Early Purple Orchid; *2. Lythrum salicaria*, Purple Loosestrife). In the first sense, XVII century, in Shakespeare's *Hamlet* 1603, among the flowers in the garlands carried by Ophelia before she drowned herself. The 'grosser name' given by 'liberal shepherds' would have been one of several which were current for orchids, such as *cullions*, *cods*, *ballocks*, in reference to the pale pair of tubers, which also occasioned the name *Dead Men's Fingers*, used, Shakespeare says, by 'our cold maids'. In the second sense, XIX century, first recorded from John Clare's *The Village Minstrel* 1821, several poems of which mention *L. salicaria* under this name, e.g. 'The Wild-flower Nosegay':

> When on the water op'd the lily buds
> And fine long purples shadow'd in the lake.

LOOFAH (the fibrous 'vegetable sponge' which forms the inside of the fruit of *Luffa aegyptica* var. *cylindrica*; and the plant). XIX century. From the Arabic of the Nile valley *lūfah*.

LOOSESTRIFE (*1. Lysimachia vulgaris*, Yellow Loosestrife. *2. Lythrum salicaria*, Purple Loosestrife). XVI century. Coined by Turner 1548 for both plants, translating the Latin *lysimachia* of Pliny, and Greek *lusimakheios* of Dioscorides (which according to Dioscorides has either red or yellow flowers), from Greek *lusis*, 'deliverance from', + '*makhe*', 'strife', 'battle'. Pliny's *lysimachia* (*L. salicaria*) is described as being so powerful 'that if placed on the yoke of inharmonious oxen it will restrain their quarrelling'. Pliny also says that his plant was discovered by Lysimachus (c. 360–281 BC, companion of Alexander and king of Thrace).

LOP-GRASS (the grass *Bromus mollis*). XVIII century, i.e. 'lob-grass'. From its loose, clumsy panicle and soft habit. Cf. *lob*, *lubber*, a lout, a lie-about.

LOQUAT (fruit of the Chinese and Japanese tree *Eriobotrya japonica*; and the tree). XIX century. Adaptation of the Cantonese *lu-kywit*, 'lu-orange', from the orange colour of the fruit.

LORDS-AND-LADIES (*Arum maculatum*, Cuckoo-pint). XVIII century. From the erect spadix sheathed in the spathe (cf. other male–female names for this plant, *Bulls and Cows*, *Men and Women*, *Stallions and Mares*; and *Cuckoo-pint(le)*, *Priest's-pintle*). As the *drakontion mikron* of Dioscorides, it was held to be aphrodisiac. 'It is uretical also, and stirrs up affections to conjunction being dranck with wine' (Goodyer's translation of Dioscorides, 1655).

LOTUS (*1*. The delicious languor-inducing fruit eaten by the Lotophagoi in Homer's *Odyssey*. *2*. The water-lily of the Nile, *Nymphaea lotus*). XVI century. Latin *lotus*, from Greek *lotos*, from Hebrew *lōt*.

LOUSEWORT (*Pedicularis sylvatica*). XVI century, Gerard 1597, translating the German *Läusekraut*, so called 'bycause the cattell that pasture where plentie of this grasse groweth, become full of lice' (Lyte 1578).

LOVAGE (*1*. The garden herb *Levisticum officinale*, native of Southern Europe, introduced in the Middle Ages. *2*. *Ligusticum scoticum*). In the first sense, XIV century. From French *livesche*, from Latin *livistica* (neuter plural as feminine), from Late Latin *levisticum*, earlier *ligusticum*, from *Ligusticus*, 'Ligurian' ('Ligusticum grows wild in the mountains of its native Liguria,' Pliny), from Greek *ligustikon*. In the second sense, XVIII century.

LOVE-APPLE (the Tomato, fruit of *Lycopersicon esculentum*; and the plant, native of western South America, introduced in the 16th century). XVI century, Lyte 1578, translating the herbalists' Latin *poma amoris* or the French *pomme d'amour* (as if tomatoes were held to be aphrodisiac).

128

LOVE-IN-A-MIST (*Nigella damascena*, native of Southern Europe introduced in the 16th century). xix century. The mist is the involucre of finely divided bracts surrounding the flower, which gave *N. damascena* the earlier recorded names of *Fennel-flower* and *Devil-in-a-Bush*. Cf. German names *Grëtli im Busch*, 'Gretel in the Bush', *Jongfer em Grönen*, 'Girl in the Green', etc., French *Cheveux de Vénus*. The sexual hair was evidently in mind.

LOVE-IN-IDLENESS (*Viola tricolor*, Pansy). xvi century, Lyte 1578. So in Shakespeare's early plays *The Taming of the Shrew* and *A Midsummer Night's Dream*, in which Love-in-idleness is a plant of desire. (Cf. 'Enameld Pansies, vs'd at Nuptials still,' in Chapman's *Ovids Banquet of Sense* 1595.) *-in-idleness* may have had the old sense of 'in vain', i.e. strong unsatisfied love or desire.

LOVE-LIES-BLEEDING (*Amaranthus caudatus*, native of the Tropics, introduced in the 16th century). xvii century, *Love lies a bleeding*. Conceptually the name is no doubt related to the *Floramor* of the German *Hortus Sanitatis* 1485; from the Latin name *flos amoris*, 'flower of love', in which *amoris* may be a garbling of *amarantus*. And cf. the 16th century German name *Blümle von der Lieb*. (Lyte 1578 writes as if *A. caudatus* was still not very familiar in English gardens, using for it the English name 'floure Gentill', which the *Oxford English Dictionary* also records for 1561.)

LUCERNE (*Medicago sativa*, Alfalfa, native of the Mediterranean region and Western Asia, introduced in the late Middle Ages). xvii century, *La Lucerne*. From the French *luzerne*, from the modern Provençal *luzerna*, *la lucerna*, 'fire-fly' (Latin *lucerna*, 'lamp'), because of the glitter or sparkling of the seeds, a name arising from the seed-trade.

LUNGWORT (*Pulmonaria officinalis*, native of Central and Northern Europe and the Caucasus, introduced in the 16th century). A lungwort is a plant good for disease or affliction of the lungs, a *pulmonaria* (*herba*) or *pulmonis herba*. The Old English *lungenwyrt* appears to have been the Hawkweed (*Hieraceum murorum L.*), the *pulmonaria gallica*, 'French pulmonaria', of 16th century medicine. The lungwort of 16th century Europe was in general *P*.

officinalis (as above), the spotted leaves of which suggested the lungs.

L U P I N (forage plants of the Mediterranean region, *Lupinus albus*, etc.; now generally the name given to the garden species from western North America, in particular forms of *L. polyphyllus*, introduced in 1826). x I v century. From Old French *lupin*, from Latin *lupinum, lupinus*, from *lupus*, 'wolf', as if *pisum lupinum*, 'wolf pea', i.e. a plant producing peas only fit for wolves.

L Y M E   G R A S S (*Elymus arenarius*). x v I I I century. An Englishing of the generic name *Elymus* (Linnaeus 1748) from the Greek *elumos*, used by Dioscorides for a kind of millet.

15. Kinds of Love-in-a-Mist (Besler, *Hortus Eystettensis*, 1613)

M

MACE (aril of the seed of *Myristica fragrans*, Nutmeg, native of Molucca). XIV century, *macys*. From the Old French *macis*, from Latin *macir*, which Pliny describes as the red bark imported from India of the large root of a tree called *macir* (Greek *makir*).

MADDER (the dye-plant *Rubia tinctorum*, native of the Mediterranean region, the roots of which were much imported into medieval England; the plant was also cultivated). Old English *mædere*, *mæddre*, of which there are other Germanic cognates.

MADONNA LILY (*Lilium candidum*, native of Southern Europe and S.W. Asia, introduced in the Middle Ages). XIX century. Symbol of the Virgin Mary, the Madonna.

MAGNOLIA (species of the genus *Magnolia*, natives of North America and Asia). XVIII century. English use of botanical Latin name (1737) in honour of Pierre Magnol (1638–1715), professor of botany and medicine at Montpellier, where he directed the botanical garden.

MAHOGANY (timber of *Swietenia mahogani*, native of the West Indies and Florida; and the tree). XVII century. A name of unknown origin, probably from Jamaica.

MAIDENHAIR (-FERN) (*Adiantum capillus-veneris*). XV century. I.e. the hair between the legs; a name suggested by the medieval Latin of the apothecaries, *capillus veneris*, 'hair of Venus', a name for this medicinal fern found in the *Herbarius* of the 4th century AD ascribed to Apuleius. Cf. the German *Frawenhär* (15th century).

MAIDENHAIR TREE (*Gingko biloba*, native of China, introduced from Japan in 1754). XVIII century. From the resemblance of the leaves to the fronds of the Maidenhair-fern.

MAIDEN PINK (*Dianthus deltoides*). XVIII century. A translation of the botanists' Latin *caryophyllus virgineus* used by Gerard (1597).

MAIZE (*Zea mays*, Indian Corn, of American origin, introduced in the 16th century). XVI century. From the Spanish *maiz*, adopted from *mahiz* in the language of the Taino people of the West Indies.

MALE FERN (*Dryopteris filix-mas*). XVI century, Turner 1562, translating the *filix mas* of the 16th century herbalists. Fuchs in his *De Historia Stirpium* (1542): '*Duo filicis sunt genera, mas scillicet et foemina*' – 'There are two kinds of fern, viz. male and female.' He goes on to refer to the Greek name *thelupteris*, 'female fern' (which Dioscorides recommends for procuring abortion in women). A female fern demanded its antithesis, a male fern, hence the *filix mas*. The female fern was taken to be Bracken (*Pteridium aquilinum*), not the Lady-fern (*Athyrium filix-femina*) of later botanical usage.

MALLOW (*Malva sylvestris*; and other species of the genus *Malva*). Old English *mealwe*, from Latin *malva*.

MALT (barley steeped and softened for brewing). Old English *malt, mealt*, with related words in most other Germanic languages, from a Germanic base meaning 'soft'.

MANDARIN (fruit of *Citrus nobilis* var. *deliciosa*, Tangerine, native of Cochin-China). XIX century. From French *mandarine*, for Spanish (*naranja*) *mandarina*, 'mandarin orange', variously explained as an orange the colour of the robes of mandarins, or an orange eaten by mandarins, or an orange which looks like a mandarin.

MANDRAKE (*Mandragora officinarum*, native of Southern Europe, grown in herbalists' gardens in the 16th century; the medicinal root was an earlier import). XIV century, in the forms *mandragge*, *mandrage* (from the Dutch *mandrage*, from the medieval Latin *mandragora*) and *mandrake*, which is apparently an assimilation of *mandrage* to *man* (describing the forked root, of magical narcotic property, often pictured as in the shape of a man) and *drake*, i.e.

dragon, as if *mandrake* were a man-shaped kind of *Dracunculus vulgaris* (the Dragon Arum), the plant which was known to medieval medicine-men as *dracontea* (*dragancia* or dragans), the *dracontium* of Pliny, and *drakontion* of Dioscorides. In medieval and Tudor times *mandragora* and *mandragoras* were also common names for the plant; which in Latin was *mandragoras*, as in Greek. The Greek *mandragoras* may derive from *namtar ira*, the Sumerian for mandrake.

MANGELWURZEL (*Beta vulgaris* ssp. *vulgaris*; the plant and the root). Late XVIII century. From German *Mangold*, 'beet', + *wurzel*, 'root'.

MANGO (fruit of the Indian tree *Mangifera indica*). XVII century. From the Dutch *mango*, from Malayan *mangga*, from the Tamil *mān-kāy*, 'mango (-tree) fruit'. (The XVI century forms *mangas*, *manga*, came from the Portuguese derivative *manga*.)

MANGOLD. *See Mangelwurzel.*

MANGOSTEEN (fruit of the Malayan tree *Garcinia mangostana*). XVI century. From the Malayan *manggistan*.

MANIOC (*Manihot esculenta*, Cassava, native of Brazil). XVII century (XVI century, *manihot*). From the Tupi-Guarani *manioca*.

MAPLE (*Acer campestre*). Old English *mapel*(*-trēow*), *mapulder*, cognate with the German *Mapeldorn*, and *Massholder*, from Old High German *mazzaltra*, from a Germanic **matlu*, 'food', the young leaves having been eaten as a pickle.

MARGUERITE (*Chrysanthemum leucanthemum*, Oxeye Daisy; and *C. frutescens*, introduced from the Canaries, 1699). XIX century. From the French *marguerite*, daisy, i.e. the Common Daisy (*Bellis perennis*), from Latin *margarita*, 'pearl'.

MARIGOLD (*Calendula officinalis*, native of the Mediterranean region, introduced in the Middle Ages). XIV century. Mary (the Virgin Mary) + *gold*, Old English *golde* (= *Chrysanthemum segetum*, Corn Marigold), i.e. the good kind of Gold (which was

16. Mandrake (Matthiolus, *Commentarii*, 1565)

used in medicine) distinguished as a flower of the Virgin from the useless Golds of farm land.

MARJORAM (*Majorana hortensis*, Sweet Marjoram, of North African origin, introduced in the Middle Ages). xiv century. From Old French *maiorane*, from medieval Latin *maiorana*. As *Wild Marjoram* applied in the 16th century to *Origanum vulgare*.

MARMALADE (confection usually of Seville Oranges). xvi century. From French *marmelade*, from Portuguese *marmeláda*, 'the pulp of quinces boiled into a consistence with sugar', from *marmélo*, a quince, from Latin *melimelum*, from Greek *melimelon*, variety of apple grafted on a quince, 'honey apple', *meli*, 'honey', + *melon*, 'apple' – which was perhaps a folk-etymologizing of a Semitic loan-word akin to Accadian *marmahu*, quince.

MARRAM (GRASS) (*Ammophila arenaria*). xvii century. From Old Norse *marálmr* (*marr* + *hálmr*), 'sea stem', 'sea reed'. In German this grass is *helm*, in Frisian *helm*, *halm*, in Dutch *helm*, in Swedish *marhalm*.

MARROW. See *Vegetable Marrow*.

MARTAGON (LILY) (*Lilium martagon*, Turk's Cap, native of the European mountains, probably introduced at the end of the 16th century). xvi century (cf. Lyte 1578, 'the Italians call it *Martagon*'), but earlier for other plants. A curious name unlikely to be derived, as the *Oxford English Dictionary* claims, from a Turkish word for a kind of turban. Martagon, possibly coined to express a relationship with the planet Mars, appears to have been used by alchemists for several plants, or for a class of Martian plants, useful in alchemy. The name is found as early as the 13th century. The 15th century Stockholm Medical MS. writes of 'Herba Martina', which is 'an herbe men clepe mortagon', evidently not describing *L. martagon*. Parallel manuscripts of the same herbal have Herba Martis, 'herb of Mars', instead of Herba Martina. Turner 1548 gives Martagon for *Listera ovata*, the Twayblade. Martagon is used for *L. martagon* in a Venetian manuscript herbal (British Museum) of the early 15th century.

In the following century the Italian herbalist Pierandrea Mattioli (1501–1577) wrote that alchemists held the Martagon Lily in high esteem for the transmutation of metals into gold. Evidently the power of this particular 'Martagon' lay in the golden-coloured root; one of the common German names of Martagon Lily, *Goldwurz*, 'gold-root', recorded as early as the 14th century, was probably descriptive both of the root colour and the alchemical use of the root.

MARSH-MALLOW (*Althaea officinalis*). Old English *mersc-mealwe*. See *Mallow*.

MARSH-MARIGOLD (*Caltha palustris*, Kingcup). Old English *mersc-mēargealle*, 'marsh horse-gall', gall in the sense of blister, swelling, bleb, describing the tight buds. Cf. the common dialect names for *C. palustris*, Mare-blob, Horse-blob. (By improbable guesswork the Old English *mersc-mēargealle* has been identified with the rare Marsh Gentian *Gentiana pneumonanthe*.) The form *marsh-marigold* first appears in the 16th century (Lyte 1578), **maregall* having been assimilated to the more familiar *marigold*, a change helped by *C. palustris* and the Marigold both having yellow flowers. See *Marigold*, and *Corn-marigold*.

MARVEL OF PERU (*Mirabilis jalapa*, native of tropical America, introduced at the end of the 16th century). XVI century, Gerard 1597, translating the Spanish *Maravilla del Peru*, in reference to the plant's extremely abundant production of scented, evening-opening flowers, and their variation; and its introduction from Peru.

MAST (the fallen fruit of woodland trees). Old English *mæst*, one of those words in the West Germanic languages of ancient ancestry suggestive of the common practices of a vanished life, in this case the communal pasturing of pigs in the autumn in the forests of oak and beech.

MASTERWORT (*Peucedanum ostruthium*). XVI century, Turner 1548, translating the German *Meisterwurz*, 'master root', from the medieval Latin name *magistrantia*. The hot root of the plant was used against 'cold' diseases.

MASTIC (gum from the Mediterranean tree *Pistacia lentiscus*). XIV century. From the Old French *mastich*, from Late Latin *mastichum*, from Greek *mastikhe*, perhaps connected with *mastax*, the 'mouth', 'jaws', and *masasthai*, 'to chew', since mastic was (see Dioscorides) and is chewed to sweeten the breath, and was an ingredient in tooth-powders.

MATÉ (tea from leaves of the tree *Ilex paraguariensis*, native of Brazil, etc.). XVIII century. By way of American Spanish *mate* from the Quechua *mati*. The word in Quechua (the language of the Bolivian and Peruvian Andes, and of the Inca empire) means 'gourd', i.e. the gourd in which the Indians prepare the tea.

MAY (flowers and leafage of *Crataegus oxyacanthoides* and *C. monogyna*, Hawthorn, Whitethorn). XVI century. Blossoming branches of hawthorn were brought home on the morning of May Day. So *May-tree*, 19th century.

MAYWEED (*Anthemis cotula*). XV century, *maydewede*, Old English *mægthe* (which survives as *Maythe*, *Maythen*), apparently akin to *mægth*, 'a maid', as if in reference to the pretty white flowers of *A. cotula* (which was hated as a corn weed for blistering the reaper's skin).

MAY WEED (blades of species of the *Laminaria* seaweeds). XIX century. The blades are cast off in the spring and washed ashore.

MAZZARD (*Prunus avium*, used as fruit-stocks; varieties of black cherry). XVI century. (Lyte 1578, mazar.) Of uncertain origin.

MEADOWSWEET (*Filipendula ulmaria*). XVI century. The meaning is not – however appropriate – 'sweet plant of the meadow', but plant used in sweetening or flavouring mead. It was earlier *Meadsweet* (15th century); cf. German *Mädesüss*; and English *Meadwort*, Old English *medowyrt* (*medu*, 'mead', + *wyrt*, 'plant'); also Danish and Norwegian *mjødurt*.

MEAL (coarsely ground cereal grains). Old English *melu*, with related words in most other Germanic languages. The meaning is that which is milled or ground, from an Indoeuropean base meaning to grind or rub (i.e. on a saddle-quern, before the invention of the rotary quern or the water-mill).

MEDICK (species of the genus *Medicago*). XV century (for Lucerne, *M. sativa*). From Latin *medica* (*herba*), Greek *medike* (*poa*), 'Median plant', 'Median grass'. Pliny's explanation is that *medica*, i.e. Lucerne, was 'foreign even to Greece, having been imported by the Medes at the time of the Persian wars under Darius'. See also *Lucerne*.

MEDLAR (*Mespilus germanica*, native of S.W. Asia and S.E. Europe, anciently introduced; the tree and the fruit). XIV century. From Old French *medlier*, medlar-tree, from *medle*, *mesle*, 'medlar', + the suffix *-ier*, from Late Latin *mespila*, Greek *mespilon*. See *Openarse*.

MELANCHOLY THISTLE (*Cirsium heterophyllum*). XVII century, Culpeper 1649. In reference to the hanging flower heads. Culpeper took this as the 'signature' of the plant, writing that the decoction of the Melancholy Thistle in wine 'expels superfluous Melancholy out of the Body', making a man 'as merry as a Cricket'.

MELILOT (species of the genus *Melilotus*, introduced; especially *M. officinalis*). XV century. From the Old French *mélilot*, from Latin and Greek *melilotos*, 'honey-clover' (*meli*, 'honey', + *lotos*, 'fodder', 'clover'), from the honey-scent of the flowers.

MELON (*Cucumis melo*, native of Southern Asia; the fruit and the plant). XIV century. From the Old French *melon*, from Late Latin *melo* (genitive *melonis*), shortened from *melopepo*, from Greek *melopepon* (*melon*, 'apple', + *pepon*, 'gourd', 'melon').

MERCURY (*Chenopodium bonus-henricus*, Good King Henry, native of Europe and West Asia, anciently introduced). XV century. From Latin (*herba*) *mercurialis*, 'herb of the god Mer-

139

cury'; this was the name of *Mercurialis annua*, Annual Mercury, with which *C. bonus-henricus* was confused. See *Dog's Mercury*.

MERMAID'S TRAIN (the seaweed *Cladophora rupestris*). XIX century. From the soft grass-like or cloth-like colonies.

MERMAID'S TRESSES (the seaweed *Chorda filum*, Sea Lace). XIX century. In reference to the string-like fronds. Also known as *Mermaid's Fishlines* (XIX century).

MEU (*Meum athamanticum*). XVI century, Turner 1548, recording it as the apothecaries' name. From Latin *meum*, from Greek *meon*, a plant described by Dioscorides.

MEZEREON (*Daphne mezereon*). XV century. English use of the medieval Latin *mezereon*, taken from the Arabic name of the plant, *māzarjūn* (borrowed from Persian), in the *Canon of Medicine* by the physician Ibn Sīnā, or Avicenna (980–1037), the Latin translation of which was the basic textbook of European medicine in the Middle Ages.

MICHAELMAS DAISY (species of the genus *Aster*, natives of North America; especially *A. paniculatus* introduced c. 1640). XVIII century. From flowering around Michaelmas (29 September). The earlier name was Starwort.

MIDSUMMER MEN (*Sedum telephium*, Orpine). XVII century. John Aubrey in his *Remaines of Gentilisme and Judaism* 1686–1687 recalls seeing cook-maids and dairy-maids sticking up 'Midsommer-men, which are slips of Orpins' in chinks of the joists – 'by paires, sc: one for a man, the other for such a mayd his sweet-heart, and according as the Orpin did incline to, or recline from the other, that there would be love or aversion; if either did wither, death.'

MIGNONETTE (*Reseda odorata*, native of North Africa, introduced in 1752). XVIII century. From French *mignonette d'Égypte*, 'little darling of Egypt'. (In France *mignonette* is commonly applied to Black Medic, *Medicago lupulina*.)

MILFOIL (*Achillea millefolium*, Yarrow). XIII century. From the Old French *milfoil*, from the Latin name (Pliny) *millefolium*, 'thousand leaf', descriptive of the minute division of the leaves.

MILK-THISTLE (*Silybum marianus*, Lady's Thistle; and *Sonchus oleraceus*, Sow-thistle). XV century for *S. marianus*, in reference to the milky veins on the leaves, and the legend that these were caused by milk which fell from the Virgin Mary's breast when she suckled Jesus. XVI century for *Sonchus oleraceus*, in reference to the milky juice (both kinds of Milk-thistle were held to increase the flow of milk in nursing mothers).

MILKWEED (species of the genus *Asclepias*, natives of North America). XIX century. In England, a name for the Sow-thistle (*Sonchus oleraceus*) recorded for the 18th century.

MILKWORT (*Polygala vulgaris*). XVI century, Lyte 1578. It was taken to be the *polygala* of Pliny and the *polugalon* (Greek *polus* 'much', + *gala*, 'milk') of Dioscorides. An infusion of the plant was given to increase mother's milk.

MILLET (grains of *Panicum miliaceum*, native of the East Indies). XIV century. From the Old French *millet*, diminutive of *mil*, from the Latin *milium*, akin to *meline*, the Greek name for millet.

MIMOSA (species of the genus *Mimosa*, especially *M. pudica*, the Sensitive Plant, native of Brazil, introduced 1638; also – particularly in flower shops – for the flowers and leafage of various Australian species of *Acacia*, or Wattle). XVIII century, in the first sense. English use of the botanists' *mimosa* (1678), Latinized from the French name *herbe mimose*, given to *M. pudica* in A. Colin's *Traité des drugues*, 1619; *mimose* seems to have been based on Latin *mimus*, an actor in a mime, as if the Sensitive Plant were silently miming a part when its leaves shrink away. In the second sense, Australian usage of late 19th century.

MIND-YOUR-OWN-BUSINESS (*Helxine soleirolii*, native of Corsica, Sardinia and the Balearic Islands, found in Corsica c. 1820). XX century. Perhaps from the self-effacing nature of the plant, creeping around in corners or hanging its tiny leaves and stems from a pot.

MINT (species of the genus *Mentha*, especially the Spearmint, *M. spicata*). Old English *minte*, in its various forms a name common to the West Germanic languages, from Latin *menta*, from Greek *minthe*.

MIRABELLE (French varieties of plum allied to bullaces and damsons). XVIII century. From French *mirabelle*, from *myrobolan*, a plum like the Myrobalan (q.v.).

MISTLETOE (*Viscum album*). Old English *misteltān* (*mistel*, 'mistletoe', + *tān*, 'twig'). *mistel* is the West and North Germanic name for *V. album*, seemingly derived from Old High German (and German) *mist*, 'dung', in reference to the way mistletoe is spread from tree to tree by birds who eat the berries and deposit the seeds on branches in their droppings. This was observed of the missel-thrush (i.e. mistletoe thrush) in the *Historia Animalium* of Aristotle (and missel-thrushes are given to nesting in apple-trees, which are the commonest host of *V. album*).

MOCK ORANGE (species of the genus *Philadelphus*, Syringa; especially *P. coronarius*, native of Southern Europe and S.W. Asia, introduced at the close of the 16th century). XVIII century. Because of the resemblance of the sweet-scented flowers to orange-blossom.

MOLY (*Allium moly*, native of Southern Europe, introduced at the close of the 16th century). In this exact sense, XVII century (Parkinson 1629). From Latin *moly*, from Greek *molu* (Homer). For the Homeric *molu*, which Hermes gave to Odysseus to combat the magical powers of Circe, *moly* was used in the 16th century (Golding's *Ovid*; Chapman's *Odyssey*, etc.). It is also used by Lyte 1578 for the *molu* of Dioscorides and the *moly* of Pliny, about the identity of which Lyte was unsure, though with Pliny he assumed it was at any rate a kind of *Allium*. See also *Amaranth*.

MONARCH (variety of pear). XIX century. A pear raised by the great horticulturist Thomas Andrew Knight (1749–1838) in

1830, the year of the accession of William IV, after whom it was named.

MONEYWORT (*Lysimachia nummularia*). XVI century, Lyte 1578. Name suggested by the German *Pfennigkraut*, 'penny plant', or by the apothecaries' Latin *nummularia* (*herba*), adjective from *nummulus*, diminutive of *nummus*, 'a coin', in reference to the coin-shaped leaves.

MONKEY-FLOWER (species of the genus *Mimulus*, of which the first were introduced from North America in the 18th century; especially *M. luteus*, introduced from Chile 1826). XVIII century. From the grinning face-like flower (cf. the expression 'monkey-face'), a name no doubt suggested by the Linnaean name *mimulus* 'little actor', diminutive of Latin *mimus*, 'actor in a farce or mime'.

MONKEY-NUT (underground fruit of *Arachis hypogaea*, native of Brazil, Groundnut, Peanut). XIX century. As a nut suitably given to monkeys in a zoo.

MONKEY-PUZZLE (*Araucaria araucana*, native of Chile, introduced 1796). XIX century. Said to have arisen from a remark made by the Benthamite lawyer Charles Austin (1799–1874) during the ceremonial planting of an *A. araucana* in the gardens at Pencarrow in Cornwall, in 1834. Austin carelessly handled the young tree and 'feelingly remarked "It would be a puzzle for a monkey"' (quoted in E. Thurstan's *British and Foreign Trees in Cornwall*, 1930).

MONKSHOOD (*Aconitum napellus*). XVI century, Lyte 1578 ('The flowers be as little hoodes'), translating Dutch *munckes capkens*. Cf. German *Münchskapffen* (1551).

MONK'S RHUBARB (*Rumex patientia*, Patience, native of Eastern Europe and Asia Minor, medicinal plant introduced in late Middle Ages). XVI century, translating the apothecaries' Latin *Rhabarbarum monachorum*. See also *Rhubarb*.

MONTBRETIA (now generally applied to garden forms resulting from a cross made in the 1870s at Nancy in France between two species of *Crocosmia* – formerly known as *Montbretia* – from South Africa, *C. pottsii* and *C. aurea*). Late XIX century. English use of the botanists' Latin *Montbretia* given in honour of the official botanist who accompanied Napoleon's Egyptian expedition in 1798, Antoine-François-Ernest Coquebert de Montbret (1781–1801). The first species of *Montbretia* were introduced from the Cape early in the 19th century.

MONTEREY PINE (*Pinus radiata*, native of California, introduced 1833). XIX century, from Monterey (bay, county and city in California, south of San Francisco), where *P. radiata* was discovered by the Scottish plant collector and botanist David Douglas (1798–1834).

MONTHLY ROSE (*Rosa damascena* var. *semperflorens*, native of the Mediterranean region, introduced in the early 17th century). XVII century, John Evelyn in his *Kalendarium Hortense* 1664, 'Monthly Rose-tree', translating the botanists' Latin *Rosa omnium Calendarum*, 'rose of all the months'.

MOON DAISY (*Chrysanthemum leucanthemum*, Ox-eye Daisy). XIX century, Anne Pratt 1861.

MOONLIGHT (*Anthriscus sylvestris*, Cow Parsley). XX century. From blanching the side of lanes, roads, etc., like moonlight.

MOONWORT (the fern *Botrychium lunaria*). XVI century, Lyte 1578, translating *Monkraut*, the German equivalent for the medieval Latin *Lunaria*; the fern was a lunar herb because the pinnae of the fronds are shaped like crescent moons. It was believed to wax and wane with the moon; and was used in alchemical experiment. 'I am happy to say nothing about the foolish behaviour of the alchemists, who make much use of this plant' (Fuchs, *De historia stirpium* 1542). *B. lunaria* was *lunaria minor*, the lesser moon-plant, in contrast to *lunaria major*, the greater moon-plant, which was *Lunaria annua*, Honesty.

MOREL (fungi of the genus *Morchella*; especially *M. esculenta*). XVII century. From the 'French *morille*, probably from a medieval Latin **mauricula*, feminine diminutive of *maurus*, 'a Moor', 'little Moor', in reference to the yellowish brown of the cap.

MORELLO (black, somewhat acid varieties of Cherry with coloured juice). XVII century. Probably from the Italian adjective *morello*, 'blackish'. (Morello has also been explained as a derivative of Flemish *marelle*, German *amarelle*, medieval Latin *amarellum*, 'little bitter one', from Latin *amarus*, 'bitter'; but the Amarelle cherries are distinguished as a group by colourless juice, and are red or yellow.)

MORNING GLORY (*Ipomœa purpurea*, native of tropical America, introduced about 1629). XIX century. An American name, but possibly a transference of the use of Morning Glory (Somerset, etc.) for the English bindweeds, *Convolvulus arvensis* and *Calystegia sepium*. *I. purpurea* was for a long while known as Indian Bindweed.

MOSCHATEL (*Adoxa moschatellina*). XVIII century. From the botanists' Latin name *moschatella*, feminine diminutive from Latin *muscus*, Greek *moskos*, 'musk', 'little musky one'. When it is raining, or when they are wet, the flowers and the whole diminutive plant smell of musk.

MOSS (plants of the bryophytic class *Musci*). Old English *mos*, related to *mēos*, both words having the collective meaning of the vegetation growing in a 'moss' or bog, and the bog itself. This and identical or similar cognates in other Germanic languages go back, like Latin *muscus*, 'moss', to an Indoeuropean **meus*, from a base *meu*, 'damp'.

MOSS ROSE (forms of *Rosa centifolia* var. *muscosa*). XVIII century. From the botanists' description of this variety of the Cabbage Rose as *Rosa Provincialis spinosissima, pedunculo muscoso*, 'very thorny Provence rose, with mossy peduncle'. The pedicels and calyx are mossy.

MOTHER OF THOUSANDS (*Cymbalaria muralis*, native of Southern Europe, introduced 1618). xix century. Also 'Mother of Millions'.

MOTHERWORT (*Leonurus cardiaca*, native of Europe, introduced in the late Middle Ages). In this sense, xvi century, Turner 1548; earlier for *Artemesia vulgaris*, Mugwort (xv century). A motherwort is a plant for the 'mother', i.e. the womb; and for hysteria, also known as the 'mother' (and thought to be caused by the rising or swelling of the womb).

MOUNTAIN ASH (*Sorbus aucuparia*, Rowan). xvi century, Gerard 1597, translating the botanists' Latin *montana fraxinus*.

MOUSE-EAR (*Hieracium pilosella*). xiii century, *musere* (the hairiness of the small leaves), translating medieval Latin *auricula muris*, translating in turn the Greek *muos ota* (Dioscorides). Later applied to *Cerastium vulgatum*, Lyte 1578.

MOUSETAIL (*Myosurus minimus*). xvi century, Lyte 1578, translating the herbalists' Latin *cauda murina*, or *cauda muris*. The name was recorded earlier for *Sedum acre*, Wall-pepper (Turner 1548).

MUGWORT (*Artemesia vulgaris*). Old English *mucg-wyrt*, *mug-wyrt*, 'midge' (Old Saxon *muggia*) + *wort*, 'plant'. A name which the English brought with them from their continental home, in German *Muggert*, as in English (though the common German name is *Beifuss*), in Schleswig *muckert*, in East Friesland *Miggert*. The name has been explained by the German use of the mugwort for catching and killing insects. Bunches of mugwort are hung up, they attract thousands of insects, and are then quickly put into a bag, and the insects are beaten to death.

MUNG, MOONG (the bean *Phaseolus aureus*, Green Gram, of East Indian origin). xix century. Hindi *mung*, related to the Accadian *mangu*.

MULBERRY (fruit of *Morus nigra*, native of Western Asia, introduced in the Middle Ages, and the tree; *Morus alba*, from

146

China, introduced late 16th century). XIV century, from Old English *mōrberie*, 'berry', + *mōr*, from Latin *morum*, from Greek *moron*, 'a mulberry' (fruit of *M. nigra*), cognate with Armenian *mor*, 'blackberry'. Cf. Old High German *mōrberi*, German *Maulbeer*.

MULLEIN (species of the genus *Verbascum*, especially *V. thapsus*). XV century. From Old French *moleine*, from *mol* (*mou*), 'soft' (from Latin *mollis*), in reference to the strikingly soft, felted leaves.

MURLIN, MURLINS (the edible seaweed *Alaria esculenta*, Dabberlocks). XIX century. From the Gaelic *muirlinn*.

MUSHROOM (any fungus with cap and stipe: now the commonly eaten Field Mushroom and the cultivated Mushroom, in contrast to Toadstool, i.e. to a fungus popularly regarded as inedible). XV century. From Old French *moisseron*, earlier *meisseron*, from the 6th century Latinized form *mussirio*, accusative *mussirionem*, of a Frankish word used in Northern France. It is found in a short treatise on food written by the Byzantine physician Anthimus, early in the 6th century, for his master Theodoric, King of the Ostrogoths. Anthimus writes that while all fungi are heavy and indigestible, the better kinds are 'mussiriones' and truffles.

MUSK (*Mimulus moschatus*, native of western North America, introduced in 1826). XIX century. In reference to the scent of musk (which is no longer perceptible in the plants of *M. moschatus* growing in England).

MUSK ROSE (*Rosa moschata*, of unknown origin, introduced in the 16th century). XVI century. A translation of the French *rose musquée*, the true *R. moschata* having a musky scent.

MUSTARD (condiment for eating with meat, made from seeds of *Brassica nigra* and *Sinapis alba*; and the plants). In the first sense, XIII century; in the second sense, XIV century. From the Old French *moustarde*, *mostarde*, seeds of mustard ground up with *moust de vin*, i.e. with 'must', new wine, from Latin *mustum*.

MYROBALAN (fruit of species of the genus *Terminalia*, natives of India and Malaya). XVI century. From French *myrobolan*, from Latin *myrobalanum*, Greek *myrobalanos*, 'unguent fruit', 'unguent date'.

MYRRH (gum-resin from the African and Arabian shrub *Commiphora myrrh*). Old English *myrre*, from Latin *murra*, from Greek *murra*, from Accadian *murru*, 'bitter'.

MYRTLE (*Myrtus communis*, native of the Mediterranean region, introduced in the 16th century). XVI century. From the medieval Latin *myrtillus*, diminutive of Latin *murtus*, *myrtus*, from Greek *myrtos*, a Semitic loan-word from Accadian *murru*, 'bitter' (see *Myrrh*). The berries, sold by apothecaries as *myrtilli*, were used in medicine.

N

NAKED LADY (*Colchicum autumnale*, Autumn Crocus). XVII century. There are similar names, descriptive of the smoothness of the flowers, naked, and unaccompanied by leaves, in other European languages, e.g. German *Nackende Jungfer*, *Nackend-hure* (naked whore), Dutch *naakte dame*.

NAPOLEON (CLOVER) (*Trifolium incarnatum*, Crimson Clover, native of Mediterranean region, introduced late in the 16th century). XIX century. Possibly a corruption of *trifolium*, but an apt name for the upstanding military appearance and brilliance of the flowers in a crop.

NARCISSUS (garden species and hybrids of the genus *Narcissus*, especially *N. poeticus*, native of Mediterranean Europe). XVI century, Turner 1548. From the Latin *narcissus*, Greek *narkissos*. One associates the name with Narcissus who pined away for love of his own reflection in the water, after which his body vanished, leaving, in Ovid's account, 'A yellow flour with milke white leaves new sprong upon the ground'. Pliny (from whom Turner took the name) wrote that it derived, 'not from the boy in the myth', but from *narke*, 'stupor' (cf. 'narcotic'), since *narcissus* had the power of 'burdening the head'.

NARD. See *Spikenard*.

NASTURTIUM (garden kinds of the genus *Tropaeolum*, of which the first was introduced from Peru late in the 16th century). XVIII century. Shortened from the botanists' Latin *Nasturtium indicum*, 'Indian Cress'. Nasturtium, from Latin *nasturcium*, name of a cress anciently, and correctly, explained as 'nose-twister', 'nose-torturer' (*nasus*, 'nose', *torquere*, 'to twist', or 'torture').

NAVELWORT (*Umbilicus rupestris*, Pennywort). xv century. Describing the depressed navel shape of the leaf. The 5th century Latin name for the plant was *umbilicus Veneris*, 'Venus's navel'; an earlier Greek name – less polite, suggesting the hairless smooth statues – was *kepos Aphrodites*, 'Aphrodite's privates' (literally 'garden').

NAVEW (*Brassica rapa*, Turnip). xvi century. From dialectical French *naveau*, French *navet*, from Old French *nef*, from Latin *napus*, 'turnip'.

NECTARINE (glabrous form of peach; and a peach-tree, *Prunus persica*, producing such fruit). xvii century. A nectarine fruit, i.e. tasting as sweet as nectar, the drink of the gods.

NENUPHAR (various species of Water-lily; especially water-lilies in debased poetry). xvi century. From medieval Latin *nenufar*, from Arabic and Persian *nīnūfar*, from Sanskrit *nīlōtpala*, *nīl*, 'blue', + *utpala*, 'water-lily', 'lotus' (i.e. the blue-flowered Indian Lotus, *Nymphaea stellata*).

NETTLE (*Urtica dioica, Urtica urens*). Old English *netele*, cognates of which are found in other Germanic languages. Referable to a Germanic **natilon*, from **naton*, from an Indoeuropean base **ned-*, to twist (the Common Nettle having been one of the ancient textile plants).

NIGHTSHADE (*Atropa belladonna*, Deadly Nightshade; *Solanum nigrum*, Black Nightshade; *Solanum dulcamara*, Woody Nightshade). Old English *nihtscada* (= *Atropa belladonna*), night + shade?, as if in reference to the black poisonous berries and growth of the plant in shady places. See *Deadly Nightshade, Dwale*.

NIPPLEWORT (*Lapsana communis*). xvii century. Coined by Parkinson 1648 as an equivalent of *pappilaris* (Latin *papilla*, a nipple), the name used by German apothecaries, since the plant was applied to cracked nipples and ulcerated breasts.

NORFOLK ISLAND PINE (*Araucaria excelsa*, from the Pacific, introduced 1796, as a pot plant). xix century. The tree is a native of Norfolk Island, to the east of Australia and north-west of New Zealand.

NORWAY SPRUCE (*Picea abies*, from Northern and Central Europe, introduced early in the 17th century). XVIII century. The spruce imported from Norway. See *Spruce*.

NOSTOC (the alga *Nostoc commune*, Falling Stars). XVII century. The name was coined by Paracelsus as an alchemical term for 'a kind of fire', and was later applied to the jelly-like masses of *N. commune* which occur on damp ground, and were supposed to have come to earth from shooting stars.

NOTTINGHAM CATCHFLY (*Silene nutans*). XIX century. From Thomas Willisel's discovery of *S. nutans* on the ruins of Nottingham Castle, in the 17th century. See *Catchfly*.

NUT-TREE (*Corylus avellana*). XIV century. Nut (Old English *hnutu*) + tree.

NUTMEG (hard seed of *Myristica fragrans*, native of the East Indies). XIV century. *Nut* + *mugue*, from an Anglo-Norman *nois mugue, variant of Old French *nois muguede*, from Provençal *notz muscada*, from medieval Latin *nux muscata*, 'musk-scented nut'.

NUX VOMICA (poisonous seed of the Indian tree *Strychnos nux-vomica*, the source of strychnine). XVI century. Medieval Latin *nux vomica* (Latin *nux*, 'nut', + *vomica*, femine of *vomicus*, 'emetic').

O

OAK (trees of the genus *Quercus*). Old English *āc*. The other Germanic languages have cognate names.

OAK-APPLE (gall caused by a gall-wasp on oak-trees). XV century. From the apple shape and rosy colour.

OAK-FERN (*Thelypteris dryopteris*). In this sense XIX century, for a fern which does not grow on oak trees; but it was coined by Turner 1548 for *Polypodium vulgare*, which does grow on oaks, as a translation of the Greek *druopteris* of Dioscorides.

OARWEED (seaweeds of the genus *Laminaria*). XVI century. Old English *wār*, 'seaweed', + weed. In Frisian dialect *wier*.

OAT (*Avena sativa*; and its edible grain). Old English *āte*. Among the Germanic languages the name is found only in English.

OCA (*Oxalis tuberosa*, native of Peru; and its edible tubers). XVII century. Spanish *oca*, from *oka* in the Quichua language of Peru.

OKRA (the tropical African *Hibiscus esculentus*, Lady's Fingers; Gumbo, and its edible fruits). XVIII century. From a West African word, perhaps *nkurama*, by which *H. esculentus* is known in the Twi language, on the Gold Coast.

OLD MAN (*Artemesia abrotanum*, native of S.E. Europe, a medicinal plant introduced early in the Middle Ages). XIX century. From the hoary leaves.

OLD MAN'S BEARD (*Clematis vitalba*, Traveller's Joy). XVIII century. From the whitish-grey feathery achenes.

OLEANDER (*Nerium oleander*, native of the Mediterranean region, Rose-bay). XVI century, Turner 1548. From the medi-

17. Oleander (Besler, *Hortus Eystettensis*, 1613)

eval Latin *oleander, oliandrum, lauriendrum.* Medieval garbling seems to have produced the beautiful name for a beautiful shrub. The Greek names for *N. oleander* were (Dioscorides) *rhododaphne,* 'rose-bay' (as if a Sweet Bay, *Laurus nobilis,* with rosy flowers), and *rhododendron,* 'rose-tree'. The medieval Latin *lauriendrum* appears to combine the *daphne* (= *laurus*) element of *rhododaphne,* and the *dendron* ('tree') element of *rhododendron.*

OLIVE (*Olea europaea;* and fruits). XIII century. From the Old French *olive,* from Provençal *oliva,* from Latin *oliva,* 'olive', 'olive tree', from Greek *eleia,* the tree which produces *elaion,* '(olive) oil'.

ONION (*Allium cepa;* and bulb). XIV century. From the Anglo-Norman *union,* from Old French *oignon,* from the Latin *unio,* accusative *unionem,* which the agricultural writer Columella, in the 1st century AD, records as a country synonym for *caepa,* the usual word for onion. *Unio* had the commoner meaning of a pearl unique in size and quality – whence 'union' in English, as in the last scene of *Hamlet*:

> The King shal drinke to *Hamlet's* better breath,
> And in the Cup an union shal he throw.

It was presumably the pearliness of peeled onions which suggested the country name. Cf. the English 'pearl-onion', small onions for pickling.

OPENARSE (*Mespilus germanica;* and fruit). Old English *openears.* For centuries the usual name, from the brown colour and shape of the fruits and their employment *against* looseness. Gradually and politely it was superseded by the French-derived *medlar.* This is one of the plant names, *apenars, apenärseken,* Frisian *iepen éarske,* found also in the ancestral lands of the English in the north of Germany.

OPIUM (juice of *Papaver somniferum,* the Opium Poppy, native of Europe and Asia). XIV century. From the Latin *opium,* from Greek *opion,* 'opium or poppy juice', diminutive of *opos,* 'plant juice or sap'.

OPOPONAX (gum from the umbelliferous plant *Opoponax chironium*, native of Southern Europe). XV century. Via Latin, the immediate source of this strange-seeming word is Greek *opoponax*, apparently *opos*, 'plant juice', + *panax*, 'all-healer', 'panacea'. But this may have been Greek folk-etymology for a loan-word akin to Accadian *kanaktu*, which had the same meaning. (The virtues of opoponax were supposedly revealed to Hercules by the centaur Chiron, so it was also called *Hercules' All heal* [Gerard 1597, etc.]. The 16th century herbalists followed Dioscorides in recommending opoponax for many conditions, including cramp, paralysis, sciatica, malaria, dropsy, asthma, cough, boils, toothache, bruises, wounds, bites and colic.) The name Opoponax is now given to other plants, especially the shrub with sweet-smelling flowers, *Acacia farnesiana*.

ORACHE (*Atriplex hortensis*, of Asiatic origin, introduced in the late Middle Ages). XV century. The Greek name for *A. hortensis* was *atraphaxus*, adapted in Latin as *atriplex*, accusative *atriplicem*; whence Gallo-Roman *atripica*, Old French *aroche*, Anglo-Norman *arasche*, and 15th century English *arage*.

ORANGE (*Citrus sinensis*, native of China or Cochin-China, and other species of the genus *Citrus*; fruit and tree). XIV century. From the Old French *orenge*. The name begins in the Dravidian language of the Malabar coast in India, *nārayam*, 'perfume within'; whence *nāran-kaj*, 'narayam fruit', adapted in Sanskrit as *nārangah*, giving Persian *nārang*, then Arabic *nārandj*, source of Italian *arancia*, and *melarancia*, 'orange-apple', which was translated in French as *pume orenge*. The initial *o* in French was caused by confusing *orenge* (*pomme d'orenge*) with the city of Orange in Vaucluse, where oranges were grown and despatched to other parts of France.

ORCHID (species of the various genera of the Orchid Family). XIX century, earlier (Turner 1548) *orchis*. From Latin *orchis*, from Greek *orkhis* (literally 'testicle'), by which kinds of orchid were known and recommended (Dioscorides, etc.) as aphrodisiac, since the stem rises erect from a pair of testicle-like tubers.

Orchid, which fits the tongue better than the older *orchis*, was shortened from the botanical Latin *Orchidaceae*, the Orchid Family.

ORCHIL (dyestuff from the lichen *Rocella tinctoria*, which was harvested in the Middle Ages from the rocks of the Mediterranean). XV century. Old French *orcheil*, from Catalan *orxella*, which in turn is from the Mozarabic *orchella*.

OREGON GRAPE (*Mahonia nervosa*, native of North America, introduced in 1820). XIX century. From its occurrence in the state of Oregon.

ORLEANS REINETTE. See *Reinette, Rennet*.

ORPINE (*Sedum telephium*, Livelong). XIV century. From French *orpin*, literally yellow, the yellow of the pigment *orpin* or *orpiment*, though *S. telephium* has magenta flowers. Medieval herborists and apothecaries recognized two kinds of *crassula*, or stonecrop – *crassula major*, which was *Sedum telephium*. and *crassula minor*, the little *Sedum acre* of walls, etc., Wall Pepper, with orpiment-coloured flowers, which gave the name *orpin* to all the stonecrops – including *S. telephium*. In French *S. telephium* remains *orpin reprise* or *grand orpin*; *S. acre* is *orpin âcre*.

ORRIS (-ROOT) (perfume from the powdered rhizomes of *Iris florentina*). XVI century. A euphoniously altered form of *Iris* (q.v.).

OSIER (species of the genus *Salix* with pliant shoots used for baskets, etc., especially *Salix viminalis*). XIV century. From Old French *osier*, *osiere*, 'that which grows in an *ausarium*', *auseria*, 'osier-bed', a Gallo-Roman word perhaps derived from a Gaulish **auesā*, 'the bed of a river' (the natural habitat of osiers).

OSMUNDA (*Osmunda regalis*, Royal Fern). XIX century. English use of the botanical Latin for the genus, which in turn was taken from the medieval Latin *osmunda*, a fern name which has never been convincingly explained. Since other ferns became known as Osmunda, *O. regalis* was distinguished both as Osmund Royal

18. Orange (Matthiolus, *Commentarii*, 1565)

(16th century), a name also given to the Male Fern, and Osmund the Waterman (Lyte 1578, *O. regalis* being a fern of bogs and streams). Since *osmunda* is first known from English medieval sources, it may be connected in some way with the English personal name Osmund.

OSWEGO TEA (*Monarda didyma*, from North America, introduced 1745). XVIII century. 'J. Bartram gathered seed of it at Oswego, on Lake Ontario, from whom it is called Oswego Tea by the people of New York; is not unpleasant' (Peter Collinson, in *Hortus Collinsonianus*, 1843).

OX-EYE DAISY (*Chrysanthemum leucanthemum*). XVIII century. As Oxeye (also = *C. segetum*), 17th century (earlier, 15th–16th century, for various ox-eyed composite flowers including species of *Buphthalmum* and *Adonis*). Translation of the medieval Latin *oculus bovis*, synonym of *buphthalmus*, from the Greek *bouphthalmon* (*bous*, 'ox', + *ophthalmos*, 'eye'), which in Dioscorides was a composite with yellow flowers.

OXFORD RAGWORT (*Senecio squalidus*, native of the volcanic cinders of Vesuvius and Etna). XX century. From the occurrence of *S. squalidus* on the high college walls of Oxford, on which it was first noticed in 1794 as an escape from the Oxford Botanic Garden.

OXLIP (*1*. The hybrid of Cowslip and Primrose, *Primula veris* × *vulgaris*. *2. Primula elatior*). Old English *oxanslyppe* 'ox's slime', i.e. ox-pat, as if growing where a pat had fallen, distinguishing the taller stouter hybrid from the more delicate *cūslyppe*, the Cowslip. (Cf., as if he was well aware of the difference, Shakespeare's 'bold oxlips' in Perdita's speech in Act IV of *The Winter's Tale*.) In sense two, 1842, after the discovery of *P. elatior* at Bardfield, in Essex. See *Cowslip*.

OXTONGUE (*Picris echioides*). XV century. An Englishing of French *langue de boeuf*, or Latin *lingua bovis* which was an adaptation of the Greek *bouglosson*, in Dioscorides one of the genus *Anchusa*. Other plants with rough tongue-shaped leaves were also known as *lingua bovis* and *Oxtongue*, including Bugloss and Borage.

OYSTER MUSHROOM (*Pleurotus ostreatus*). XIX century. From the shape; suggested by Latin *ostreatus*, pertaining to oysters.

OYSTER PLANT (*Mertensia maritima*). XIX century. The leaves have a flavour like oysters.

OYSTER THIEF (the seaweed *Colpomenia peregrina*, immigrant to English and French Atlantic coasts early in this century). XX century. Adaptation of French *Voleuse d'huîtres*, referring to the way plants of *C. peregrina* grew on oysters in the Breton oyster-beds, and lifted them by means of their airfilled thalli.

P

Paigle (*Primula veris*, Cowslip). xvi century, *paggle*. Of obscure origin, possibly connected with obsolete verb to 'paggle', as of a cow's neck, to hang and shake, i.e. paggling, loosely hanging flowers.

Pak-choi (*Brassica sinensis*, Chinése Cabbage, introduced 1770). xix century. Cantonese form of *pai ts'ai*, 'white vegetable', from the whitish stems and ribs.

Palm (trees of the family *Palmaceae*). Old English *palm*, from the Latin *palma*, with the same meaning. (Literally *palma*, 'palm of the hand', from the hand-shaped or palm-shaped leaves of many kinds.)

Palma Christi (*Ricinus communis*, Castor Oil Plant, of African origin; introduced in the 16th century). xvi century, Turner 1548. The medieval Latin name, meaning 'Christ's palm', 'Christ's hand', in reference to the palmately divided leaves.

Pampas-grass (*Cortaderia selloana*, native of the Argentine, introduced in 1843). xix century. Growing on the pampas or plains of the Argentine (*pampas*, from Spanish *pampa*, plural *pampas*, from the Quechua *pampa*, 'a plain').

Pandanus (trees of the genus *Pandanus*, natives of the East Indies, Polynesia, etc.). xix century. English use of the botanists' Latin, from the Malay name *pandan*.

Pansy (*Viola tricolor* var. *hortensis*). xvi century, *pensee*; late xvi century, *pansy*. From the French name *pensée*, literally a 'thought', i.e. the flower as a symbol of remembrance, especially of the giver.

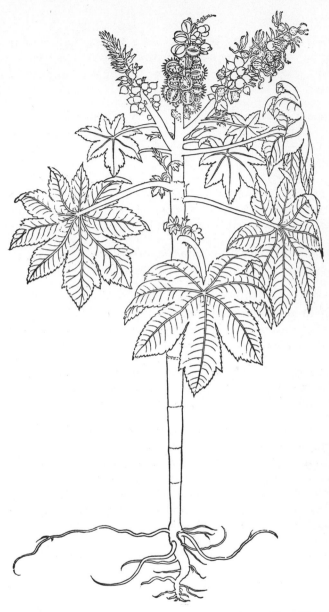

19. Palma Christi, Castor Oil Plant (Brunfels, *Herbarum Vivae Eicones*, 1530)

PAPAW, PAWPAW (fruit of *Caprica papaya*, native of tropical America; and the tree). XVI century, *papaios* (the fruit); XVII century *papawe*, *papaw-tree*, *papa*. From the Spanish *papaya*, adapted from the Carib name.

PAPRIKA (Hungarian red pepper from fruits of *Capsicum frutescens*, a native of the Tropics). XIX century. Hungarian *paprika*, from Serbian *paprika*, diminutive of *papar*, from modern Greek *piperi*, pepper.

PAPYRUS (the sedge *Cyperus papyrus*, from Africa, Syria and Southern Europe). XIV century. Latin *papyrus*, from Greek *papuras*, possibly a loan-word from Egyptian (hence also *paper*).

PARASOL MUSHROOM (*Lepiota procera*). XIX century. Perhaps a translation of the German *Parasolschwamm*.

PARSLEY (*Petroselinum crispum*, from Southern Europe, introduced in the Middle Ages). XIV century. From Old French *perresil*, from popular Latin *petrosilium*, from Latin *petroselinum*, from the Greek *petroselinon* (*petros*, 'rock', + *selinon*, 'celery').

PARSLEY FERN (*Cryptogramma crispa*). XVIII century. Describing the tufted leaves, with curled edges, resembling curled forms of parsley.

PARSLEY-PIERT (*Aphanes arvensis*). XVI century, Gerard 1597. A garbling of French *percepier*, *perce-pierre*, a plant which grows in rocky places, piercing (*percer*, 'to pierce, bore') the rock or stone, and so used in medicine against the stone. The Flemish botanist Mathias de l'Obel had found *A. arvensis* at Bristol, and had described it (1574) as the '*Percepier Anglorum*', '*perce-pierre* of the English'.

PARSNIP (*Pastinaca sativa*). XIV century, *passenep*. From the Old French *pasnaie*, from Latin *pastinaca*, 'parsnip', 'carrot', connected with the verb *pastinare*, 'to dig'. The English derivative of *pasnaie* acquired its ending in -*nep* by assimilation with *nep* (Old English *nēp*, *nǣp*), 'turnip' (q.v.).

PASQUE FLOWER (*Anemone pulsatilla*). XVI century, first as *Passeflower* (Lyte 1578), from French *passe-fleur*, which Lyte seems to explain by writing that the root of anemones boiled in *passum* (Latin for raisin-wine) was good for the eyes (a remedy from Dioscorides). Gerard 1597 changed *passeflower* to *Pasque Flower*, 'They flower for the most part about Easter, which hath mooved me to name it Pasque flower, or Easter flower' (pasque, paske, from Old French *pasques*, Easter).

PASSION-FLOWER (species of the genus *Passiflora*, native of tropical America, first introduced in the 17th century). XVII century. A translation of the botanists' Latin *flos passionis*. The crown of thorns, the apostles (ten of them, less Peter and Judas) and the signs of Christ's passion were detected in the parts of the complex flower.

PASSION-FRUIT (fruit of *Passiflora edulis*, native of Brazil, introduced from the West Indies). XVII century. See *Passion-flower*.

PATIENCE (*Rumex patientia*, Monk's Rhubarb, native of Eastern Europe and Asia Minor, a medicinal plant introduced in the late Middle Ages). XV century. From French *patience* (*lapatience*), an assimilation to *patience*, in the ordinary sense, of *lapacion*, from Latin *lapathium*, *lapathum*, from Greek *lapathon*, with the same meaning.

PATCHOULI (the mint *Pogostemon heyneanus*, native of Malaysia; and the scent prepared from it). XIX century. From Hindi *pacholi*, from the Tamil *paccilai*, 'green leaf'.

PAULOWNIA (the tree *Paulownia tomentosa*, native of China, introduced 1840). XIX century. From the botanists' Latinization (1837) of Paulovna, the tree having been named *Paulownia imperialis* in honour of Anna Paulovna, daughter of the mad Tsar Paul I of Russia, and consort of William II, King of the Netherlands (after whom one of the Dutch polders is also named Anna Paulovna).

PEA (seeds of *Pisum sativum*, Garden Pea, native probably of Western Asia; and the plant). XVII century. A new singular formed from *pease*, Old English *pise*, from the Late Latin *pisa*, from Latin *pisum*, from Greek *pison*.

PEACH (fruit of *Prunus persica*, of Chinese origin; and the tree). XIV century, *peche*. From the Old French *pesche*, from Low Latin *persica*, from Latin *persicum* (*malum*), 'Persian apple'.

PEACOCK'S-TAIL (the seaweed *Padina pavonia*). XIX century. Describing the fan-shaped frond.

PEANUT (underground fruit of *Archis hypogaea*, of Brazilian origin; and the plant, Ground-nut, Monkey-nut). XIX century. Descriptive of the podded nuts.

PEAR (fruit of *Pyrus communis*, probably anciently introduced; and the tree). Old English *peru*, *pere*, 'a pear', from West Germanic **pera*, from Low Latin *pira*, a feminine singular use of Latin *pira*, plural of *pirum*, 'a pear'.

PEARLWORT (species of the genus *Sagina*). XVII century, John Ray 1660. *pearl* + *wort* (q.v.), from the small capsule, or the unopened flowers.

PEARMAIN (a medieval variety of winter pear; later a variety of apple). XV century, in the earlier sense (though on record in a Latinized form in the 13th century). XVI century as an apple (Gerard 1597). From the Old French *permain*, *parmain*, said to derive from Romanic **Parmanus*, 'Parman (fruit)', i.e. from Parma.

PECAN (nut of *Carya pecan*, native of the southern United States and Mexico; and the tree). XVIII century, *paccan*, from the name in one of the Algonquin languages.

PELARGONIUM (species of the genus *Pelargonium*, 'Geranium', from South Africa, of which many were introduced during the 18th century; also garden forms). XIX century. English use of the Latinized botanical name (1787), from Greek *pelargos*, 'a stork'. The name indicates the relationship of the genus to the similar

20. Pasque Flower (Brunfels, *Herbarum Vivae Eicones*, 1530)

true Geraniums, in the same family of *Geraniaceae* (Greek *geranion*, from *geranos*, 'a crane').

PELLITORY-OF-THE-WALL (*Parietaria diffusa*). XVI century, Turner 1548. *P. diffusa* was so named to distinguish it from the original Pellitory, Pellitory-of-Spain, the composite *Anacyclus pyrethrum*, of S.E. Europe (16th century). 14th century *peletre*, from Old French *peletre*, *piretre*, from Latin *pyrethrum*, from Greek *purethron* (*puretos*, fever). Pellitory-of-the-wall was known in the 14th century as *paritarie*, from the Old French *paritaire*, from the Latin (*herba*) *parietaria*, plant of walls (*paries*, 'a wall'). Likeness of sound combined the two names into Pellitory.

PENNYROYAL (*Mentha pulegium*). XVI century. A garbling of 16th century *puliole real* (in medieval Latin *pulogium regale*), from Old French *pouillol* (*réal*), (royal) *pouillol*, from Latin *puleium*, *pulegium*, as if 'flea-plant': according to Pliny the flea (Latin *pulex*) is killed by the scent of the burning leaves.

PENNYWORT (*Umbilicus rupestris*). XV century. 'This herbe hath lewys lyk a peny' – Stockholm Medical MS (c. 1440).

PENSTEMON (species of the genus *Penstemon*, natives of North America, the first of which were introduced in 1758). XIX century. English alteration of the botanists' Latin, earlier *pentstemon*, a combination of Greek *pente*, 'five', and *stemon*, a 'thread', the penstemons having four fertile and one barren stamen.

PEONY (species of the genus *Paeonia*). Old English *peonie*, from the Latin *paeonia*, Greek *paionia*, the paionian plant, i.e. the plant of Paion, the physician of the gods. Pliny: 'The most anciently discovered herb is the Peony, which still keeps its discoverer's name.' On the authority of Dioscorides *P. corallina* and *P. officinalis*, held to be male and female forms of the same plant, were highly regarded in medieval and 16th century medicine.

PEPPER (dried berries of the vine *Piper nigrum*, native of Assam and Burma). Old English *pipor*. This and other West Germanic forms derive from Latin *piper*, going back via Greek *peperi* to Sanskrit *pippalī*. The spice spread anciently both to the West and to the Far East. Black pepper (the dried berries which turn black): Old English, *blacum pipore*. White Pepper (berries with the black pericarp removed): 14th century. See also *Red Pepper*.

PEPPER DULSE (the seaweed *Laurencia pinnatifida*). XVIII century. From its pungency of taste and smell.

PEPPERMINT (*Mentha × piperita*). XVII century, John Ray 1696, coined by him after the discovery of the hybrid in Hertfordshire. He described it, unrealistically, as a mint *sapore fervido Piperis*, 'with a fiery taste of pepper'.

PERIWINKLE (*Vinca minor*, and *V. major*, Lesser and Greater Periwinkle). XIV century, *pervenke*, *parvink*, *pervinke*, from the Old French *pervenche*, from Latin *vinca pervinca*, *vinca pervica*, a plant mentioned by Pliny. The altered form appears early in the 16th century, *perwyncle*.

PERRY (fermented pear juice). XIV century, *pereye*, from Old French *peré* (modern French *poiré*), from a Late Latin form *peratum*, 'that which was made from pears', from Latin *pirum*. See *Pear*.

PERSIMMON (fruit of species of the genus *Diospyros*, especially *D. kaki*, native of China and Japan, and *D. virginiana*, from the eastern United States; and the tree). XVII century, first as *putchamin*, *pessemmine*, from an Indian word in one of the Algonquin languages.

PE-TSAI, PI-TSI (*Brassica pekinensis*, introduced 1770). XX century. See *Pak-choi*.

PETUNIA (species of the genus *Petunia*, introduced from the Argentine in the 1820s and 1830s; and garden forms). XIX century. English use of the botanists' Latin name, 1789, given by A. L. de Jussieu, from French *petun*, 'tobacco', from the Portuguese *petum*, from the Tupi-Guarani *petyn*. Petunias resemble species of *Nicotiana*, the Tobacco Plant; and both belong to the *Solanaceae*.

PHEASANT'S EYE (*Adonis annua*). XVIII century. Describing the scarlet, black-centred corolla. Cf. the German name (16th century) *Teufelsauge*, 'devil's eye'.

PHLOX (species of the genus *Phlox*, from North America, of which the first was introduced in 1725). XVIII century. English use of the botanical name *Phlox*, from Latin *phlox*, an unidentifiable plant mentioned by Pliny, from Greek *phlox*, flame.

PIGNUT (tuber of *Conopodium majus*, Earthnut; and the plant). XVII century. The earliest quotation of pignut in the *Oxford English Dictionary* comes from Shakespeare's *Tempest*, Caliban telling Trinculo and Stephano that with his long nails he'll dig them pignuts.

PILEWORT (*Ranunculus ficaria*, Lesser Celandine). XVI century, Lyte 1578, translating German *Feigwurtz*. From the resemblance of the root-tubers to the swellings caused by piles, or the 'fig'. *R. ficaria* was used against piles.

PILLWORT (*Pilularia globulifera*). XIX century. Botanists' invention, descriptive of the small pill-like pericarps.

PIMENTO (in English 18th century usage, and modern American usage, fruit of *Capsicum frutescens*, Red Pepper, of tropical countries. In 18th century and modern English usage, Allspice, from the dried berries of the tree *Pimenta officinalis*, native of the West Indies and Central America; also the tree). From Spanish *pimienta*, from medieval Latin *pigmenta*, plural of *pigmentum*, 'spice', 'spiced drink' (cf. the medieval English *piment*, wine with honey and spices, such as Absolon sends to the carpenter's wife Alison, in Chaucer's *The Milleres Tale*, from *pigmentum*, by way of French *piment*); from Latin *pigmentum*, in the late sense, 'juice of plants'.

PIMPERNEL (*Anagallis arvensis*, Poor Man's Weatherglass, Scarlet Pimpernel). XV century, *pympernol*, *pympernelle*, which was properly the name for the Burnet Saxifrage (*Pimpinella saxifraga*), from French *pimprenelle*, Old French *piprenelle*, from medieval Latin *pipinella*, deriving ultimately from Latin *piper*, 'pepper', owing either to the taste of the leaves of *P. saxifraga* or the resemblance of its ripe fruits to peppercorns. The transference of the name to *Anagallis arvensis* is unexplained, though it suggests that a not dissimilar name for *A. arvensis* – perhaps medieval Latin *prunella*, 'little coal', diminutive of Latin *pruna*, 'a living coal', descriptive of the scarlet flowers – was absorbed in the familiar name for *Pimpinella saxifraga*.

Scarlet Pimpernel is not found until the 19th century (Anne Pratt 1855).

PINE (trees of the genus *Pinus*). Old English *pīn*, *pīn-trēow*, 'pine-tree', from the Latin *pinus*.

PINEAPPLE (fruiting inflorescence of *Ananas comosus*, native of tropical America, introduced in the 17th century; and the plant). XVII century. From the resemblance to an extra large cone from a pine-tree, such cones having been formerly described as 'pine-apples'.

PINK (species, with garden forms, of the genus *Dianthus*). XVI century. From the resemblance of the open flowers of garden pinks to the 'pinks' or ornamental openings cut to show colour in Elizabethan dress (*pink*, the colour, derives from the flowers, not the other way round). Cf. the Tudor name *Pink-needle*, i.e. needle for pinking, for species of *Erodium*, Stork's-bill.

PIÑON (edible seeds of *Pinus cembroides* var. *edulis*, native of the western United States; and the tree). XIX century. Spanish *piñon*, 'a pine-nut'; from Spanish *pina*, from Latin *pinea*, 'a pine cone' (*pinus*, 'a pine').

PIPEWORT (*Eriocaulon septangulare*). XIX century. From the jointed roots.

PIPPIN (varieties of apple). XV century. Apple trees raised from pippins or seeds. (XVIII century *pip* is a shortened form of *pippin*.)

PISSABED (*Taraxacum officinale*, Dandelion). XVI century, Gerard 1597. Perhaps a translation of the French *pissenlit* (first recorded 1536 – cf. Dutch *pisse-bed, Pis in't bed*), though the name has too wide a currency to be derived from Gerard's *Herbal*. *T. officinale* has genuine diuretic powers.

PISTACHIO (kernel of the fruit of *Pistacia vera*, native of the Mediterranean region and the East; and the tree). XVI century. From the Italian *pistacchio*, from the Latin *pistacium*, from Greek *pistakion*, 'pistachio nut' (Greek *pistake*, 'pistachio tree').

PITCHER-PLANT (species of the genus *Nepenthes*, natives of East Indies, Ceylon and South China, of which the first was introduced in 1789; and the genus *Sarracenia*, from North America, of which *S. purpurea* was introduced in 1640). XIX century. From the pitcher-shaped leaves.

PITTOSPORUM (species of the genus *Pittosporum*, many of them from Australia and New Zealand). XIX century. English use of botanists' Latinized name coined by Sir Joseph Banks, who accompanied Captain Cook on his New Zealand–Australian voyage (1769–1770). From Greek *pitta*, pitch, and *sporos*, seed, from the glutinous pulp round the seeds.

PLANE (trees of the genus *Platanus*). XIV century. From the Old French *plane*, from Latin *platanus*, from Greek *platanos* (*P. orientalis*), equivalent to 'broad tree', from Greek *platus*, 'broad'. See also *London Plane*.

PLANTAIN (species of the genus *Plantago*). XIV century. From the Old French *plantain*, from Latin *plantaginem*, accusative of *plantago*, 'a plantain', from *planta*, 'sole of the foot', because of the flat spread of the leaves.

PLOUGHMAN'S SPIKENARD (*Inula conyza*). XVI century, Gerard 1597. A spikenard, because of its fragrant root, for ploughmen. See *Spikenard*.

PLUM (fruit of *Prunus domestica*; and the tree). Old English *plūme*, from Low Latin *pruna*, neuter plural of Latin *prunum*, 'a plum', taken as a feminine singular.

POINSETTIA (*Euphorbia pulcherrima*, native of Mexico and tropical America, introduced 1834). XIX century. English use of the botanists' Latin (1836) *Poinsettia*, the earlier generic name, in honour of the American Minister to Mexico, Joel Roberts Poinsett (1779–1851), who had discovered *E. pulcherrima* in 1828.

POISON IVY (*Rhus radicans*, native of North America). XIX century. The earliest record in the *Oxford English Dictionary* is 1857. Earlier (18th century) *Poison Vine*.

POKE, POKE-WEED (*Phytolacca americana*, American Night-shade, introduced from Virginia early in the 17th century). XVIII century. From *puccoon*, in one of the Algonquin languages of Virginia.

POLICEMAN'S HELMET (*Impatiens glandulifera*, native of the Himalayan region, introduced in 1839, established as an escape by 1855). XX century. From the shape of the flowers.

POLYANTHUS (various kinds of Primula in the hybrid group *Primula polyantha*). XVIII century. From the botanical Latin *polyanthus* (Greek *polu-*, 'many', + *anthos*, 'flower') used early in the 17th century of such kinds of primula, and in common English speech by the second and third decades of the following century.

POLYPODY (the fern *Polypodium vulgare*). XV century. From the Latin *polypodium*, Greek *polupodion*, 'many-footed', describing the branched and creeping rhizome, which was used in medicine (Dioscorides).

POMEGRANATE (*Punica granatum*, of South Asiatic origin; the fruit and the shrub). XIV century. From the Old French *pume grenate*, 'pomegranate apple', *grenate* deriving from a Low Latin form, *granata, of Latin *granatum*, 'pomegranate', originally *pomum granatum* 'apple with many seeds' (*granum*, a seed).

POMELO (*Citrus paradisi*, Grapefruit; which arose, apparently as a sport or by hybridization, in Barbados before 1809). XIX century, but first as a name in the East Indies for the large

171

Shaddock or Pompelmous (*Citrus maxima*). Possibly a shortening and alteration of the (Dutch) *Pompelmous* (q.v.).

POMPELMOUS (*Citrus maxima*, the Shaddock, of South Asiatic origin). XVII century. From Dutch *pomp*, 'a melon' + *limoes*, a Malayan borrowing from Portuguese *limão*, 'citrus', 'lemon'.

PONDWEED (species of the genus *Potamogeton*). XVI century, Lyte 1578.

POOR MAN'S WEATHER-GLASS (*Anagallis arvensis*, Scarlet Pimpernel; the seaweed *Laminaria saccharina*, Sea Belt). XIX century, probably earlier for *A. arvensis*. The *Oxford English Dictionary*'s first quotation is 1847; *weather-glass* was a common term by the mid 17th century. The first quotation given for the parallel *Shepherd's Weather-glass* comes from John Clare's *Shepherd's Calendar*, from 'May':

> Pimpernel dreading nights and showers
> Oft calld 'the shepherds weather glass'.

POPLAR (trees of the genus *Populus*). XIV century. From the Old French *pouplier*, earlier *peuple*, to which the *-ier* of French tree-names was added; from Latin *populus*.

POPPERING, POPPERING PEAR (pear varieties well known in Tudor times, and till the 18th century). XVI century. Remembered from Mercutio's bawdy punning over Romeo (*Romeo and Juliet*, II, i, 39, 40):

> O Romeo that she were, O that she were
> An open et cetera, or thou a Poprin Peare

(for 'open et cetera,' see *Open Arse*, above). From the Belgian town of Poperinghe, near Ypres. The *Oxford English Dictionary* quotes *Speake Parrot* by John Skelton (1460?–1529): 'In poperinge grew peres, whan parrot was an eg.'

POPPY (species of the genus *Papaver*; especially *P. rhoeas*, Field Poppy). Old English *popig*, earlier *popæg*, from a Low Latin form, **papavum*, of Latin *papaver*, apparently related to the Sumerian *pa pa*, 'poppy'.

PORT (dessert wine from Portugal). Late XVII century. Shortened from *O Port wine*, or *Port-wine*, wine from Oporto (*O Porto*, Portuguese 'The Port'), from which the Portuguese wines are shipped.

PORTER (dark beer or stout brewed from charred malted barley). XVIII century. Shortened from *Porter's Beer*, *Porter's Ale*, i.e. beer favoured by porters.

PORTUGAL LAUREL (*Prunus lusitanica*, native of Portugal and Spain, introduced 1648). XVIII century. Translation of the botanical Latin *laurocerasus lusitanica*.

POTATO (tuber of *Solanum tuberosum*, native of the Andes, introduced into Spain c. 1569 or earlier; into England possibly in 1586). XVI century. Originally the name of the Sweet Potato (*Ipomoea patatas*), then applied to tuber of *S. tuberosum*, Gerard 1597, writing of 'Virginia potatoes', which he distinguished from 'common potatoes'. From Spanish *patata*, adaptation of *batata*, the name in the Arawak language encountered by Columbus in Hispaniola (Haiti). See also *Sweet Potato*.

PRIEST'S PINTLE (*Arum maculatum*, Cuckoo-pint, Lords-and-Ladies). XVI century, Lyte 1578. *Pintle*, i.e. penis. Lyte was perhaps translating the *Pfaffenpint* of German 16th century herbalists, but cf. the earlier *Cuckoo-pintel*, *Cuckoo-pint*, in which the cuckoo is substituted for the parson. There are, or have been, equivalent names in several languages, e.g. French *membre d'évêque*, *vit de prêtre*, Swedish *munksvans*. Very obviously 'signed', the plant was regarded as aphrodisiac. See *Cuckoo-pint*.

PRIMROSE (*Primula vulgaris*). XV century. From the medieval Latin *prima rosa*, 'first rose', rose being used in a loose sense as flower – 'first' because the first flower to mark the spring. Primrose was applied also to *Primula veris*, the Cowslip, without clear distinction; Parkinson 1629 says expressly that he uses Primrose for *P. vulgaris*, and Cowslip for *P. veris*, to avoid this confusion.

PRIMULA (the Polyanthus species and forms in the genus *Primula*). XVIII century. From the Linnaean botanical use of the medieval Latin *primula* (*veris*), 'firstling of the spring', i.e. the Cowslip, as the name of the genus. See *Polyanthus* and *Primrose*.

PRINCE'S FEATHER (*Amaranthus hybridus* var. *hypochondriacus*, introduced from America in the late 17th century; also the Asian *Polygonum orientale*, introduced from the East Indies in 1707). XVIII century. The name was first used in the 17th century (Parkinson 1629) for the London Pride (*Saxifraga spathularis* × *umbrosa*).

PRIVET (*Ligustrum vulgare*). Old English **pryfet*, which is found in place-names, such as Privett in Hampshire, and minor names of the Middle Ages such as Prevetmore or Privetheye, though *privet* is not on record by itself until 1542. Of unknown derivation, though the suffix *-et* indicates, as in other words from trees and shrubs, a collective, i.e. a thicket of the plant named in the first syllable of *privet*.

PROVENCE ROSE (*Rosa gallica*, from Western Asia; *Rosa* × *centifolia*, hybrids of which *R. gallica* is one of the parents). XVI century, Lyte 1578 '*Roses of Province*'. Translation of French *rose de Provins*, rose grown round Provins (in Seine-et-Marne, S.E. of Paris), which was a rose-growing centre from the 13th century. *Provins*, *Province*, was assimilated to the more familiar *Provence*.

PRUNE (dried plum). XIV century. From the Old French *prune*, from the Low Latin *pruna* (see *Plum*).

PUFFBALL (fungi of the genus *Lycoperdon*). XVII century. The ball which puffs out spores. Earlier, a *puff* (16th century).

PUMPKIN (fruit of *Cucurbita pepo*, of ancient cultivation and unknown home). XVII century. Apparently an altered form of the earlier *pompion*, *pompon* (16th century), from obsolete French *pompon*, 15th century *pépon*, from Latin *pepo* (accusative *peponem*), from Greek *pepon*, 'melon', shortened from *sikuos pepon*, 'ripe sikuos', name of a kind of melon eaten when fully ripe.

PURGING FLAX (*Linum catharticum*). XVIII century. Botanists' translation of the botanical Latin. *L. catharticum* was sold as a laxative in apothecaries' shops as *Herba Lini cathartica*; Johnson in his edition of Gerard's *Herbal* (1633) describes its preparation, cooked gently in a pipkin of white wine.

PURPLE LOOSESTRIFE (*Lythrum salicaria*). XVI century, coined by Turner 1548, to distinguish *L. salicaria* from *Lysimachia vulgaris*, Yellow Loosestrife. He simply translated the herbalists' Latin *Lysimachia purpurea*. See *Loosestrife*.

PURSLANE (*Portulaca oleracea* var. *sativa*, anciently cultivated, of unknown origin. Introduced in the Middle Ages). XIV century. From the Latin name of the herb in Pliny *porcilacca* (= *portulaca*), which seems to have been assimilated to the Italian *porcellana*, 'cowrie'; hence, in the same two senses of 'cowrie' and 'purslane', Old French *porcelaine*, English *purcelan*, *purslane*. (The ceramic porcelain also descends from *porcellana*, *porcelaine* = cowrie; *porcellana* in this sense deriving from Latin *porcella*, little sow, with the meaning 'little cunt'.)

PYRACANTHA (species of the genus *Pyracantha*). XVIII century. English use of the botanical Latin *pyracantha*, in *Crataegus pyracantha* (= *Pyracantha coccinea*, native of Southern Europe, introduced early in the 17th century). From Greek *purakantha*, 'fire-thorn', a shrub mentioned by Dioscorides.

Q

QUAKING-GRASS (*Briza media*). XVI century, Gerard 1597: 'shakers, or quaking grasse'.

QUARRENDEN (apple variety producing small deep red fruit). XV century. Probably named after the parish of Quarrendon, in Buckinghamshire. Also *Devonshire Quarrenden*, 17th century. (Quarrendens are still grown in Cornwall, Somerset and Devonshire.)

QUASSIA (the tropical American tree *Quassia amara*, and its medicinal wood). XVIII century. English use of the generic Latin name coined by Linnaeus in honour of a negro slave of Surinam (Netherlands Guiana), Graman Quassi, who used an infusion of the bitter quassia bark against malaria.

QUEEN ANNE'S LACE (*Anthriscus sylvestris*, Cow Parsley). XX century. Describing the lace-like appearance of the white umbels along roads, lanes, footpaths. Probably a vague use of *Queen Anne* for something associated with the past; though it would be pleasant to suppose this is an older name referring, not to the 18th century queen, but to the beautiful Queen Anne, Danish wife of James I, and patroness of poets.

QUETSCHE (variety of plum, originating in Eastern Europe). XX century. German *Quetsche, Zwetsche*.

QUICKEN, QUICKEN-TREE (a northern name for *Sorbus aucuparia*, the Rowan, or Mountain Ash). Old English **cwicen*; 14th century, *quiken*. Cognate with Old English *cwicu*, 'possessing life', possibly with the sense that *S. aucuparia* was credited with apotropaic power against supernatural craft and enchantment.

QUILLWORT (*Isoetes lacustris*). XVIII century. From its quill-like leaves.

176

QUINCE (*Cydonia oblonga*, native of Central Asia; the tree and the fruit). XIV century. Plural, which came to be used as a singular, of *coyn*, from Old French *cooin*, from *cotoneum* (*malum*), *cydoneum* (*malum*), Latin equivalent for Greek *melon kudonion*, 'apple of Cydonia' (Canea), in Crete.

QUININE (medicinal alkaloid from the bark of species of the South American genus *Chinchona*). XIX century. Coined, to describe the alkaloid, from *quina*, Spanish spelling of the Quechua word *kina*, bark. In Quechua (the language of the empire of the Incas) the Indians spoke of the medicinal bark as *kina-kina*, from which was derived, via Spanish, the English *quinaquina* (17th century), synonymous with *Peruvian Bark* or *Jesuit's Bark*.

QUITCH (*Agropyron repens*, Couch, Twitch). A form of *Couch* (q.v.).

R

RADISH (*Raphanus sativus*, of ancient European and Asiatic cultivation). Old English *raedic*, from the Latin *radix* (accusative *radicem*), literally 'root', though *radix* also meant 'radish' to the Romans.

RAGGED ROBIN (*Lychnis flos-cuculi*). XVIII century. *Robin* (from the French), familiar diminutive of the personal name Robert; *ragged* because of the ragged appearance of the four-cleft petals. Perhaps a late medieval name which was not recorded earlier, in this sense, than 1777 (the *Oxford English Dictionary* quotes an earlier date, 1741, for *Ragged Robin* as *Lychnis coronaria*, the Dusty Miller or Rose Campion of gardens). *Robin* suggests the endearing familiarity and cheerfulness of Ragged Robin in May and June.

RAGWORT (*Senecio jacobaea*). XV century. *rag* (because of the ragged appearance of the leaves) + *wort* (q.v.).

RAISIN (dried grape). XIII century. From the Old French *resin*, from the Low Latin *racimus*, Latin *racemus*, bunch of grapes.

RAMPION (*1. Campanula rapunculus. 2. Phyteuma tenerum*, Round-headed Rampion). XVI century. Indirectly from the medieval Latin *rapuntium*, which appears to derive from Latin *rapum*, 'turnip'. In the 16th century Rampion was commonly known as *rapunculum, quasi parvum rapum*, 'as if a little turnip' (Fuchs, *De historia stirpium* 1542); and it was identified with the wild turnip of Dioscorides. The swollen root was eaten as a salad. As *Phyteuma tenerum*, 18th century – a botanical use: since *C. rapunculus* is uncommon in England (? a plant remotely of garden origin), the name Rampion was going begging.

RAMSONS (*Allium ursinum*). Old English *hramsen*, plural of *hramsa, hramse*. A very old and in its various forms widespread name for a common and conspicuous plant, the Danish and Norwegian (and German) *rams*. Not infrequent in place-names, e.g. Ramsey, in Essex.

RAPE (*Brassica napus*). XIV century. From Latin *rapa*, or *rapum*, 'turnip'.

RASPBERRY (fruit of *Rubus idaeus*; and the plant). XVII century. Raspberries were earlier *rasps, raspis, raspises* (16th century), possibly taken as a fruit from which a drink could be made like a *raspis* (15th century), a sweet red wine imported from France, a *vinum raspatum*, or wine made from *râpes*, grapes from which the seeds have been removed.

RED CURRANT (fruit of *Ribes rubrum*; and the plant). XVII century. See *Currants*.

RED HOT POKER (species of the genus *Kniphofia*, natives of South Africa, the first of which *K. uvaria* was introduced in 1707). Late XIX century. The glow of the red and yellow inflorescence.

RED PEPPER (fruit of *Capsicum frutescens*, native of the Tropics, introduced in the 16th century). XVI century. See also *Pepper, Cayenne, Chilli*.

RED RATTLE (*Pedicularis palustris*). XVI century, Lyte 1578. Because of the rattling of the seeds in the dry capsules.

REDWOOD (*Sequoia sempervirens*, native of the Pacific coast of United States, introduced in 1846). XIX century. From the colour of the bark.

REED (*Phragmites communis*, and its tall stems). Old English *hrēod*, the first element of many names of stream or riverside places, such as the (Wiltshire) Rodbourne, 'reed stream'. A word of the West Germanic languages.

REEDMACE (*Typha latifolia* and *T. angustifolia*). XVI century, Turner 1548. Because of the resemblance of the long reed-like stem, with its close-packed inflorescence, to a mace or sceptre.

REINE CLAUDE (*Prunus domestica* ssp. *italica*, Greengage, native probably of Asia Minor, and its varieties; re-introduced, from France, in 1724). XVIII century. From the French *reine-claude*, originally *prune de la reine Claude*, i.e. plum of Claude, the wife of François I, who became queen in 1515, with her husband's accession, and died in 1524, worn out by repeated childbirth. If introduced into France from Italy, in her time, the Reine Claude or Greengage would be an item of the French Renaissance along with the turreted pleasure-château of Chambord, and the elongated nudes of the artist Primaticcio.

REINETTE, RENNET (apple varieties of French origin such as the Orleans Reinette, and Reinette du Canada). XVI century. From French *rainette*, literally 'frog' (*rainette verte*, tree frog), to which apples of the *reinette* or *rainette* type were likened because of their small size and brownish markings on a yellow ground; though *reine*, queen, influenced the spelling, as if such good eaters were little queens among apples.

RESTHARROW (*Ononis repens*). XVI century. An English equivalent for the medieval Latin names *resta bovis* (French *arrête-boeuf*), 'ox stop', and *remora aratri*, 'hindrance of the plough', since the deep and strong roots were said to check the oxen and the instrument.

RHODODENDRON (species of the genus *Rhododendron*, the first of which was introduced in 1656; especially *R. ponticum*, from Spain, etc., introduced in 1763, from the garden of a convent between Cadiz and Gibraltar). XVIII century. From Greek *rhododendron*, Greek *rhodon*, 'rose', + *dendron*, 'tree', the name for the Oleander (*Nerium oleander*), in which sense *Rhododendron* was first used in English.

RHUBARB (*Rheum rhaponticum*, native of Siberia, introduced in the 16th century). XV century, as the imported drug, i.e. the root of the Chinese species *Rheum officinalis*. From the Old French *reubarbe*, from the medieval Latin *rheubarbarum*, which was influenced by the earlier name in the Latin of the Middle Ages, *rhabarbarum*, reproducing the Greek *rha barbaron* 'foreign *rha*' or rhubarb; in which *rha* is the ancient name of the Volga, along which the drug came to the Ancient World. The *rheu-* in the medieval *rheubarbarum* reproduces the other Greek name for the drug, *rheon*, Latinized as *rheum*, from the Persian *rēwend*.

Latin names for rhubarb, which is one of the oldest medicaments of Chinese civilization, were *radix pontica* and *rha ponticum*, the root, or the *rha*, which came across the Pontus or Black Sea, or, as Dioscorides wrote in his account of rhubarb, from 'beyond the Bosphorus'.

RIBSTON PIPPIN (variety of eating apple). XVIII century, first as *Ribston Park Pippin*, the original tree having been raised by Sir Henry Goodricke at Ribston Hall, near Knaresborough, in the West Riding of Yorkshire, from seed which came from the neighbourhood of Rouen and was planted in 1707. See *Pippin*.

RIBWORT (*Plantago lanceolata*). XV century. *Rib* + *wort*, 'plant', from the narrow rib-like and ribbed leaves. Earlier, in Old English, *ribbe*.

RICE (grain of the grass *Oryza sativa*, native of Indonesia). XIII century. From the Old French *ris*, from the Italian *riso*, from Low Latin **oryzum*, from Latin *oryza*, reproducing the Greek *oruza*, from the Persian *brīzi*, the origin of which is the Sanskrit *vrihih*.

ROBIN'S PINCUSHION (the Bedeguar or Briar Ball produced on Wild Rose stems by the gall wasp *Rhodites rosae*). XIX century. Descriptive of the Robin Redbreast colour of the Bedeguar accompanied by the thorns of the rose.

ROCAMBOLE (*Allium scorodoprasum*, Sand Leek). XVII century. Borrowed by gardeners from French *rocambole*, from German *Rockenbolle*; *Rocken* (*Roggen*), 'rye', + *Bolle*, 'onion'. From growing as a weed in fields of rye.

ROSA MUNDI (*Rosa gallica* var. *versicolor*, a sport of the Provence Rose, raised by the 17th century). XVIII century, Miller 1731: 'The Rose of the World, *or* Rosa Mundi' (Latin *mundus*, 'the world').

ROCK HUTCHINSIA (*Hornungia petraea*; formerly *Hutchinsia petraea*). XIX century. Botanical Latin *Hutchinsia*, given in 1812 to the genus in honour of the Irish botanist Ellen Hutchins (1785–1815), of Ballylickey, Co. Cork, a hunter after Alpine plants, mosses, etc.

ROCKET (*Eruca sativa*, native of the Mediterranean region, introduced in the 16th century). XVI century. The Latin name for this now little grown, very hot-tasting salad plant was *eruca* (Pliny, etc.); whence Italian *ruca*, diminutive *ruchetta*, French *roquette*; and so *Rocket*, in English, a name also given to *Hesperis matronalis* (Garden) Rocket, 18th century, and, qualified in various ways, to other cruciferous plants.

ROCKROSE (species of the genus *Helianthemum*). XVIII century. See also *Rose*.

ROSE (flower of species of the genus *Rosa*: and the plant). Old English *rōse*, from the Latin *rosa*, which is akin to Greek *hrodon*. Also used in names for other plants with flowers of distinctive beauty, irrespective of colour, as Rockrose, Rose of Sharon, Christmas Rose, Corn Rose (= the Poppy, *Papaver rhoeas*).

ROSE OF SHARON (*Hypericum calycinum*, native of S.E. Europe, long introduced). XIX century, for this plant. In the Bible (*Song of Solomon*), Authorized Version, 1611, *Rose of Sharon* (i.e. beautiful flower from Sharon, the coastal plain of Israel) was a charming invention by the translators, for which they had no warrant in the Hebrew. By 1878 Rose of Sharon was being applied to the garden and shrubbery favourite *Hypericum calycinum*, which was in need of a name better than the ones by which it was commonly known, *Eastern St John's Wort*, *Large-flowered St John's Wort*, and *Large-Calyxed St John's Wort*. See also *Rose*.

ROSEBAY WILLOWHERB (*Epilobium angustifolium*). XVII century. The flowers resemble those of the *Rose Bay*, Turner's name (1548) for the Oleander (*Nerium oleander*).

ROSEMARY (*Rosmarinus officinalis*, native of the Mediterranean region, introduced in the late Middle Ages). XV century, earlier *rosmarine*, from the Latin name of the plant, *rosmarinus*, *ros maris*, 'dew of the sea', explained by Pliny as a plant which grew 'in dewy places'. *ros* was assimilated to rose, and *marine* to (the Virgin) Mary.

ROSE-ROOT (*Sedum rosea*). XVI century, Gerard 1597, who, like other 16th century herbalists and like Dioscorides before them, rightly remarks on the delicious rose perfume of the root-stock. The English name translates the German *Rosenwurz* or the Latin *rhodia radix*, or the Greek *rhodea rhiza* of Dioscorides.

ROWAN-TREE (*Sorbus aucuparia*, Mountain Ash). XVI century, Turner 1548: 'rountree'. From Old Norse **raun*, whence also Norwegian *raun*.

ROYAL FERN (*Osmunda regalis*). XIX century. A translation of the Latin *regalis*, 'royal', of the botanical name, + *fern*. See *Osmunda*.

RUBBER (solidified latex of the tree *Hevea brasiliensis*, native of the Amazon region). XVIII century (1789) from the use of rubber as a rubber-out of the marks made by the lead pencil. *India rubber*, 1812 (*Indian rubber*, 1789).

RUDBECKIA (species of the North American genus *Rudbeckia*, of which *R. lacianata* was introduced in 1640). XX century. English use of the generic Latin name given in honour of the Swedish botanist Olof Rudbeck the Elder (1630–1702), founder of the botanic garden at Uppsala.

RUE (*Ruta graveolens*, native of Southern Europe, introduced in the Middle Ages). XIV century. In the Greek of the Peloponnese the name for *R. graveolens* was *hrute* (instead of the ordinary Greek *peganon*). This gave *ruta* in Latin; from which came Old

English *rūde*, which disappeared, and Old French *rue*, which entered English in the 14th century – in time for the punning association of the bitter-leaved *Rue* with *rue* and *ruth*, which so pleased Elizabethan poets.

RUNCH (used for both *Raphanus raphanistrum*, Wild Radish, White Charlock, and *Sinapis arvensis*, Charlock). XVI century. North Country and Scottish farmer's name for two damaging weeds; of unknown origin.

RUPTUREWORT (*Herniaria glabra*). XVI century, Gerard 1597, translating the *herniaria* of 16th century botanists. *H. glabra* was used as a remedy for rupture (Latin *hernia*).

RUSH (species of the genus *Juncus*). Old English *risc*, a West Germanic word, from an Indoeuropean base meaning to bind, or plait.

RUSSET (varieties of apple). XVIII century (17th century *Russetting*). From the russet or red-brown colouring.

RUSTY-BACK (the fern *Ceterach officinarum*). XIX century. From the brown scales on the underside of the frond.

RUTABAGA (*Brassica napus* var. *napobrassica*, Swede, introduced from Sweden in 1781). Early XIX century. From *rotbagga* 'ram's root' (in Swedish dialect of West Götland).

RYE (*Secala cereale*, anciently from Western Asia, and probably first raised as a crop in North Central Germany). Old English *rȳge*. There are cognate names in other Germanic languages.

RYE-GRASS (*Lolium perenne*). XVII century, *Ray-grass*, in which *ray* was assimilated to *rye* in the 18th century. From *ray* (=Darnel) + *grass*.

21. Rowan-tree, Quicken (Matthiolus, *Commentarii*, 1565)

S

SACK (white wine once imported from Spain and the Canaries). XVI century, *wyne seck*. From French *vin sec*, 'dry wine'.

SAFFLOWER (the dried flowers, a yellow and red dyestuff, of the Asiatic *Carthamus tinctorius*; and the plant). In the first sense, XVI century; in the second sense (the plant was grown in England), XVII century. From Dutch *saffloer*, from Old French *saffleur*, from Italian *saffiore*, *asfiore*, from Arabic *safrā*, feminine of *asfar*, 'yellow' (the second syllable suggesting *-fiore*, 'flower', in the Italian).

SAFFRON (dried stigma of *Crocus sativus*, native of Asia Minor; and the plant, introduced in the 16th century). In the first sense XIII century. From the Old French *safran*, from the medieval Latin *safranum*, from the Arabic *zacfarān*.

SAGE (*Salvia officinalis*, native of the Mediterranean region, introduced in the Middle Ages). XIV century, *sauge*, from the Old French *saulje*, from Latin *salvia*, 'plant of good health' (Latin *salvere*, 'to be in good health'; *salvus*).

SAGO (the Sago Palm, *Metroxylon sagu*, native of Malaysia; and the starchy substance derived from its pith). XVI century. From the Malayan *sāgū*.

SAINFOIN (*Onobrychis viciifolia*). XVII century. From French *sainfoin* (*sain*, 'sound', 'healthy', + *foin*, 'hay'), i.e. a dried crop good for cattle.

ST BARNABY'S THISTLE (*Centaurea solstitialis*, native of Southern Europe and Western Asia, an introduced weed of forage crops, etc.). XVI century. From blossoming around St Barnabas's Day, 11 July.

186

ST DABEOC'S HEATH (*Daboecia cantabrica*, growing in Ireland, in Co. Mayo and Galway). XIX century. Translation of the Latin *Erica S. Dabeoci* in John Ray's *Historia Plantarum* 1704, which in turn translated the Irish *fraoch Da-bhéog*, '(St) Dabhéog's heather', a local name reported to Ray by the discoverer of the plant, Edward Lhuyd (1660–1709). From Lhuyd, Ray added in his Latin text that 'superstitious young women carry twigs of it around with them against unchastity'.

ST GEORGE'S MUSHROOM (the edible mushroom *Tricholoma gambosum*). XIX century. From its springtime appearance round about St George's Day, 23 April.

ST JOHN'S WORT (species of the genus *Hypericum*). XVI century, Turner 1538, *Saynt Iohanns gyrs*, i.e. grass; Turner 1551, *saint Iohns grass*, or *saynt Iohns wurt*. Earlier (15th century) *Herb Jon*. Translation of the medieval Latin *herba Sancti Ioannis*, from flowering around midsummer (St John's Day, 24 June).

ST PATRICK'S CABBAGE (*Saxifraga spathularis*). XIX century. A distinctive plant of the mountain districts of Ireland, the country of St Patrick, as well as N.W. Spain.

SALAD (dish of raw lettuce, etc.). XV century. From Old French *salade*, from Old Provençal *salada*, from Romanic **salata* (*herba*), 'salted vegetable', past participle of **salare*, to salt; from Latin *sal*, salt.

SALAD BURNET (*Poterium sanguisorba*). XIX century. From the use of the young leaves in salads. Cf. John Evelyn, writing in 1693, 'Burnet . . . is a very common and ordinary sallet furniture.' See *Burnet*.

SALEP (starchy food from tubers of species of the genus *Orchis*, including *O. mascula*, Early Purple Orchid). XVII century. From the Spanish *salep*, from Arabic *sahlab*, *thaᶜlab*, shortened from *khusy ath-thaᶜlab*, 'testicles of the fox', a name for orchid species. Cf. *orchis*, from Greek *orkhis*, literally 'testicle'. Salep was regarded as strengthening and aphrodisiac. See *Orchid*.

SALLOW (various low-growing species of the genus *Salix*). Old English *salig*, a form of *salh*, 'willow', 'sallow', which has cognates in the Germanic languages, also in the Latin *salix*.

SALSIFY (*Tragopogon porrifolius*, Vegetable Oyster, native of the Mediterranean region, introduced from France and Italy into English gardens c. 1700). XVIII century. From French *salsifis*, *salsefie*, from Italian (*erba*) *salsifica*, of unknown origin.

SAMPHIRE (*Crithmum maritimum*). XVI century, Elyot 1545, *sampere*; in Shakespeare's *King Lear* 1608, *sampire*. From French (*herbe de*) *Saint-Pierre*, a name which is not current in France where *C. maritimum* is called *bacile*. The saint was evidently an intrusion into some such name as *herbe des pierres*, 'plant of the rocks', among which Samphire grows. Samphire was believed to be of use in kidney affections and in breaking stones.

SANDAL(WOOD) (perfumed heartwood of the Indian tree *Santalum album*). XIV century. From the Old French *sandal*, from the medieval Latin *sandalum*, from Arabic *sandal*, from Sanskrit *candana*, 'burning wood', i.e. wood for incense. Sandal was the perfumed wood of Buddhism, used for making statues, blending incense, etc.

SANICLE (*Sanicula europaea*). XV century. From the Old French *sanicle*, from medieval Latin *sanicula*, a diminutive formed from Latin *sanus*, 'whole', 'sound', Sanicle having been reputed as a wound herb. 'The iuyce of Sanicle dronken, doth make whole and sound all inward, and outwarde woundes and hurtes' (Lyte 1578).

SARACEN'S CONSOUND (*Senecio fluviatilis*, Broad-leaved Rag-wort, native from Central and Southern Europe to Siberia, introduced probably in the late Middle Ages). XVI century, Lyte 1578, translating the apothecaries' Latin *consolida sarrecinica*, i.e. *consolida*, 'whole-making plant', of the Saracens. Renowned as a wound herb.

SARSAPARILLA (*Smilax officinalis*, native of South America; its roots; and the drink made from the roots). XVI century, *zarza parilla, sarsa parilla.* From the Spanish *zarzaparrilla, zarza,* bramble (Arabic *sharas*), + *parilla,* diminutive of *parra,* 'a vine', 'little prickly vine'.

SASSAFRAS (the tree *Sassafras albidum,* native of Mexico and the southern United States; and its medicinal bark). XVI century. From the Spanish *sasafras,* from Latin *saxifraga (herba),* a plant which has the medicinal power of breaking up the stone, for which Sassafras was administered. See *Saxifrage.*

SATINWOOD (wood of the Indian tree *Chloroxylon swietenia;* also of the West Indian *Fagara flava*). XIX century. The wood, much used in early Victorian and mid-Victorian furniture, veneering etc., has a satin sheen.

SATSUMA (Tangerine type fruit of *Citrus nobilis* var. *unsiu*). XX century. Named after the province of Satsuma in the south of Kyushu, the southern island of Japan.

SAUCEALONE (*Alliaria petiolata,* Jack-by-the-Hedge, Garlic Mustard). XVI century. I.e. a sauce strong enough by itself. Saucealone was variously used as a spring sauce (Turner 1551), a sauce with salt fish (Gerard 1597), and a sauce with boiled mutton.

SAVIN (*Juniperus sabina,* native of Central Europe and Western Asia; and its dried tops, which were used to procure abortion). The Old English was *safine*; but *savin,* 14th century, derives from the Old French *savine.* The source is Latin (*herba*) *Sabina,* 'Sabine plant', described by Pliny.

SAVORY (*Satureja montana,* Winter Savory, native of S.E. Europe and Western Asia; and *S. hortensis,* native of the Mediterranean region. Introduced in the Middle Ages). XIV century, *saverey,* an assimilation to English *savery* (i.e. savoury, tasty) of the Old French *sarriee,* from the Latin *satureia* (i.e. *S. hortensis*), which was much used in Roman cookery.

SAVOY (form of *Brassica oleracea* var. *capitata*). XVI century (Lyte 1578). Shortened from *Savoy colewort,* translating the

Dutch *Savoyekool* or botanist's Latin *brassica sabauda*, cabbage from Savoy, the former Italian duchy in the western Alps, ceded to France in 1860. Its French name is *chou de Milan*, Milan cabbage, as if the form had crossed into the western Alps from the Lombardy plain.

SAWWORT (*Serratula tinctoria*). XVI century, Gerard 1597, translating the Latin *serratula*, a name used by Pliny as a synonym of Betony (*Stachys officinalis*), and given to *S. tinctoria* because its leaves also have a serrate, or saw-like, edge, 'somewhat snipt about the edges like a sawe' – Latin *serra* – 'whereof it tooke his name' (Gerard). Because of this saw-like 'signature' *S. tinctoria* was considered a wound-herb.

SAXIFRAGE (species of the genus *Saxifraga*). XVI century, for Meadow Saxifrage (*S. granulata*); XV century, for Burnet Saxifrage (*Pimpinella saxifraga*). From Old French *saxifrage*, from Latin (*herba*) *saxifraga* (*saxum*, 'rock', + *frangere*, 'to break'). Following Pliny, who wrote that the Maidenhair Fern was called *saxifragum* because it was 'admirable for expelling stones from the body and breaking them', it was thought that a saxifrage was a plant of any kind (often 'signed' by growing in rock-clefts or in stony or dry places) accredited with the same power, especially the Burnet Saxifrage (*Pimpinella saxifraga*), in describing which the 15th century herbal *British Museum MS. Harley 3840* says, 'Saxifrage ys an erbe that me clepyth saxfrage othyr stone-breke.' The first saxifrage in the modern sense to be so called, *S. granulata*, the Meadow Saxifrage (Turner 1568, 'White Saxifrage'), has root bulblets or granulations by which it lives through the winter. These were sold in apothecaries' shops as *Semen Saxifragae Albae*, 'seed of White Saxifrage', for making bladder and kidney medicine (Gerard 1597).

SCABIOUS (*Knautia arvensis*, Field Scabious, Gipsy Rose; also *Scabiosa columbaria*, Small Scabious). XIV century. From the apothecaries' Latin *scabiosa*, a plant for the scab or scabies and similar conditions, the more valued *scabiosa* having been *K.*

arvensis. Its use against such scabby afflictions was probably suggested by the scabby appearance of the involucral bracts.

SCAMMONY (Radix Scammoniae, medicinal root of *Convolvulus scammonia*, Purging Bindweed, native of Asia Minor; and the plant). In the first sense, xv century; in the second sense, xvi century. From the Old French *escamonée, scamonee*, from Latin *scammonea*, from Greek *skammonia*.

SCARLET PIMPERNEL. See *Pimpernel.*

SCARLET RUNNER (the Kidney-bean *Phaseolus coccineus*, native of tropical America, introduced in the early decades of the 17th century). xix century. Called in the 18th century *Scarlet-bean*, and grown at first for covering arbours.

SCORZONERA (*Scorzonera hispanica*, native of Central and South-eastern Europe, introduced in the 16th century). xvii century. English use of the herbalists' Latin *scorzonera*, from Italian *scorzonera*, 'snake plant', the plant of the *scorzone*, a kind of snake (*Elaphe longissima*, the snake of the Aesculapian temples). Scorzonera was held to be efficacious against snakebite.

SCOTCH FIR (*Pinus sylvestris*). A name first recorded in the late xvii century. *Scotch Pine* was used by Miller (1731). *Scots Fir* and *Scots Pine* are not on record until 1797, according to the *Oxford English Dictionary.*

SCOTCH PINE. *See above.*

SCOTCH THISTLE (*Onopordium acanthum*, Cotton Thistle, native of Southern and N.E. Europe, Russia and Western Asia, probably introduced; and rare in Scotland). xviii century. The *Oxford English Dictionary*'s first record of *Scotch Thistle* comes from the translation, 1705, of Abraham Cowley's *Plantarum Libri*:

> Whilst the Scotch Thistle with audacious Pride
> Taking Advantage, gores your bleeding Side.

(The thistle which the medieval Scotch would have had in mind as the heraldic emblem of their country, for its pride and prickliness in defence, would have been the common Spear Thistle, *Cirsium vulgare*, or the Marsh Thistle, *Cirsium palustre*.)

SCURVY-GRASS (*Cochlearia officinalis*). XVI century, Gerard 1597 (Turner 1568, *Scurby wede*, *Scurby wurt*). Because of the employment of *C. officinalis* against scurvy, as by Tudor seamen (on which see Gerard). *Scurvy-grass* (and Turner's names, above) were probably from the Dutch or Frisian. The antiscorbutic properties of *C. officinalis* were first mentioned by Dodoens 1553, who describes it as a specific much taken by Dutchmen and Frisians.

SEA BELT (the ribbon-fronded seaweed *Laminaria saccharina*). XVI century, Turner 1548, 'it may be named in englishe, fysshers gyrdle or sea gyrdel, or sea belte.'

SEA-BLITE (*Suaeda fruticosa* and *S. maritima*). XIX century. From Latin *blitum*, Orache, Spinach; from Greek *bliton*. Blite was used in the 15th century probably for *Chenopodium album*, Fat Hen.

SEA-BUCKTHORN (*Hippophae rhamnoides*). XVIII century. Miller 1731, *H. rhamnoides* having been taken by 16th century herbalists to be a kind of *rhamnus* or buckthorn. See also *Buckthorn*.

SEA GIRDLE (the seaweed *Laminaria saccharina*). See *Sea Belt*.

SEA HEATH (*Frankenia laevis*). XVIII century.

SEA HOLLY (*Eryngium maritinum*). XVI century, Turner 1548, 'Eryngium is named in englishe sea Hulver or sea Holly.'

SEAKALE (*Crambe maritima*). XVII century, John Evelyn, in *Acetaria. A discourse of sallets* 1699, *Sea-keele*. sea + *kale*, cabbage; a plant of coastal sand and shingle.

SEA LACE (the seaweed *Chorda filum*). XVII century. From the long lace- or cord-like fronds.

SEA LAVENDER (*Limonium vulgare*). XVI century, Gerard 1597. Descriptive of the stiff habit of the plant and its lavender-coloured flowers.

SEA LETTUCE (the seaweed *Ulva lactuca*, Green Laver). XVII century. From the green lettuce-like fronds.

SEA OAK (the seaweed *Halidrys siliquosa*). XVI century, Gerard 1597, translating the 16th century herbalists' Latin *quercus marinus*.

SEA PINK (*Armeria maritima*, Thrift, Lady's Cushion). XVIII century, Miller 1731. See also *Pink*.

SEA STOCK (*Matthiola sinuata*). XIX century. For its first record the *Oxford English Dictionary* quotes Matthew Arnold in his poem 'The Forsaken Merman', which he wrote in 1848 or 1849:

> . . . we rose through the surf in the bay.
> We went up the beach, by the sandy down
> Where the sea-stocks bloom, to the white-walled town.

SEA THONG (the seaweed *Himanthalia elongata*). XVII century, Gerard 1633. From the long fertile branches like straps or thongs.

SEAWEED (marine *Algae*). XVI century. The *Oxford English Dictionary* records *seaweed* first for 1577.

SEA WHISTLE (the seaweed *Ascophyllum nodosum*, Knotted Wrack). XIX century. The air bladders on the fronds are cut by children into whistles.

SEA-WRACK (seaweed washed up on the foreshore). XVI century, Turner 1557. See *Wrack*.

SEDGE (collectively, reed or grasslike plants of bog and marshy land; plants of the genus *Carex*). Old English *secg*, German *segge*, from an Indoeuropean base *seq-*, 'to cut', meaning plants with a serrate edge which cuts. In the second, restricted, sense, XVIII century.

SELFHEAL (*Prunella vulgaris*). XIV century. A selfheal was a wound-herb; and the name was applied to various species laid on cuts and wounds – including Sanicle (*Sanicula europaea*), Burnet Saxifrage (*Pimpinella saxifraga*) and Scarlet Pimpernel (*Anagallis arvensis*). Finally Selfheal became restricted to *P. vulgaris*.

SENNA (leaflets and pods of *Cassia acutifolia*, native of tropical Africa, and *C. angustifolia*, from India and Arabia). XVI century. English use of the medieval Latin *senna*, from the Arabic *sanā*.

SENSITIVE PLANT (*Mimosa pudica*, native of Brazil, introduced 1638). XVII century. A translation of the botanists' Latin *herba sensibilis*.

SEQUOIA (*Sequoia sempervirens*, Redwood, from the Pacific coast of the United States, introduced in 1846). XIX century. English use of the botanical Latin name in honour of Sequoyah (1770–1843), who devised a syllabary for the Cherokee language. He was a half-caste (hence the name Sequoyah, slang term for a half-caste, from the Cherokee word for opossum), properly named George Gist, son of a Cherokee woman and a German father.

SERVICE, SERVICE TREE (*Sorbus domestica*, Sorb, Sorb-tree, native of Southern Europe, Asia Minor and North Africa, introduced in the 16th century). XVI century, a use of the plural *serves*, the sorb-apples or fruits of *S. domestica*, 15th century *serve* (both the fruit and the tree), from Old English *syrfe* (*syrfetrēow*), which derives ultimately from Latin *sorbus*, Service Tree (*sorbum*, a sorb-apple).

SESAME (*Sesamum orientale*, native of the Asian tropics; the seeds and the plant). XV century. From Latin *sesamum*, *sesama*, from Greek *sesamon*, 'sesame seed', from *sesame*, the plant, a Semitic loan-word, ultimately from Accadian *samassammu*.

SEVILLE ORANGE (the sour *Citrus aurantium*, native probably of Cochin-China). Late XVI century *Civil Orange*, oranges from the neighbourhood of Seville in Spain.

SHADDOCK (*Citrus maxima*, native of Malaysia and Polynesia, ancestral to the Grapefruit, *C. paradisi*). XVII century. Said by Sir Hans Sloane (1707) to be named after a Captain Shaddock, commander of an East Indiaman, who brought the seed to Barbados while on a passage from the East Indies to England.

SHAGGY CAP, SHAGGY MANE (the fungus *Coprinus comatus*).
XIX century. Self-explanatory.

SHALLOT (*Allium ascalonicum*, the bulb and plant). XVII century.
With loss of initial *e* a form of *eschalot*, early 18th century, from
French *eschalotte*, diminutive of Old French *escaluigne*, from the
Latin *ascalonia* (*caepa*), onion from Ascalo (Ascalon in Palestine),
described at some length by Pliny, who lists the shallot, the
'*ascalonia* named from a Judaean town', as one of the kinds of
onion favoured by the Greeks.

SHAMROCK (clover leaf worn on St Patrick's Day, 17 March).
XVI century. From the Irish *seamrōg*, Little Clover, diminutive
of *seamar*, clover. Used by several 16th and 17th century writers,
copying each other, as if the Irish for watercress. So Edmund
Spenser, in his *View of the Present State of Ireland* (written c. 1597–
1598) on famine-stricken Irishmen after the wars in Munster,
'and yf they founde a plotte of water-cresses or shamrokes, there
they flocked as to a feast for the time.' Gerard 1597 had properly
interpreted Shamrock as clover, taking it to mean *Trifolium
pratense*, Meadow Clover. The usual identification is with the
small *Trifolium dubium*, Lesser Yellow Trefoil. (The legend that
St Patrick used a Shamrock leaf to explain the Three in One and
One in Three of the Trinity was well known by the 18th century;
in fact this late association with St Patrick christianized an older
use of the Shamrock as a charm against witches, fairies and the
like, who were especially active around 17 March, which was
taken to be the end of winter and the beginning of spring.)

SHEEP'S-BIT (*Jasione montana*). XVIII century. Called by Lyte
1578 *Sheepes Scabious*, i.e. the Scabious which grew on barren hills
instead of in cornfields, and was therefore cropped by sheep. He
was translating either the Dutch *schaaps-scabieuse* or the French
scabieuse de brebis. See *Scabious*.

SHEEP'S FESCUE (*Festuca ovina*). XVIII century. Translation of
the Linnaean name *Festuca ovina*. A grass of upland sheepwalks.
See also *Fescue*.

SHEPHERD'S CRESS (*Teesdalia nudicaulis*). XIX century. A botanists' name for this uncommon species, on the model of *Shepherd's Purse, Shepherd's Needle*.

SHEPHERD'S NEEDLE (*Scandix pecten-veneris*). XVI century, Lyte 1578, translating the herbalists' Latin *acus pastoris* or the French *aiguille de berger*. From the long fruits. (*Acus pastoris* had also been used for the Musk Storksbill, *Erodium moschatum*, for which *Shepherd's Needle* is found in 1562.)

SHEPHERD'S PURSE (*Capsella bursa-pastoris*). XV century. From the resemblance of the siliculae or seed cases to the bag-like purses or pouches which were carried by shepherds and others, suspended on strings from the waist.

SHEPHERD'S ROD (*Dipsacus pilosus*, Small Teasel). XVII century, Gerard 1633. A translation for this uncommon teasel of Latin *virga pastoris*, which apothecaries used for the large Common Teasel. The 16th century herbalists explain that shepherds would often use long stems of the Common Teasel for directing their flocks, stripping away the prickles at one end for a hand-hold.

SHEPHERD'S WEATHER-GLASS (*Anagallis arvensis*, Scarlet Pimpernel). XIX century. See *Poor Man's Weather-glass*.

SHERRY (the white wine from southern Spain). XVI century. Shortened from *sherris*, from Spanish *vino de Xeres*, 'wine of Xeres', i.e. the town of Jeréz de la Frontera (Latin *oppidum Caesaris*, 'Caesar's town').

SHIELD FERN (species of fern included, or formerly included, in the genus *Aspidium*; now *Polystichum lobatum*, Hard Shield Fern, and *P. setiferum*, Soft Shield Fern). XIX century. A translation of the botanical Latin *Aspidium*, from Greek *aspidion*, 'a small shield', describing the indusium, the tissue which covers each spore-container.

SHIRLEY POPPY (garden strain of the wild poppy *Papaver rhoeas*). XIX century (1886). From Shirley, ecclesiastical parish in Croydon, in the London suburbs, where the Rev. W. Wilkes raised the first Shirley poppies in his vicarage garden.

SHITTIM WOOD (timber of *Acacia nilotica*, native of North Africa, material of the ark, etc., of the Tabernacle). XVII century. Authorized Version 1611, *Exodus*, e.g. xxv, 10: 'And they shall make an ark of shittim wood.' From Hebrew *shittim*, plural of *shittāh*, probably from Egyptian *sh-n-s-t*.

SHOREWEED (*Litorella uniflora*). XVIII century. Suggested by the generic name *litorella*, 'little one of the shore' (from Latin *litus*).

SIBERIAN CRAB (*Malus baccata*, native of Siberia, Manchuria and China, introduced 1784). XIX century (1803). From the small fruits, like those of a Crab Apple.

SILVERWEED (*Potentilla anserina*). XVI century, Lyte 1578, translating the herbalists' Latin, *argentina* (*herba*), or the Dutch *zilverkruid*. Gerard 1597 says, 'The later Herbarists do call it Argentina, of the silver drops that are to be seen in the distilled water thereof when it is put into a glasse, which you shall easely see rowling and tumbling up and down in the bottom'; but the name may be simply descriptive of the silvery leaves.

SISAL (*Agave sisalana*, of unknown provenance in Central America; and its fibre). XIX century. Shortened from Sisal Hemp, Sisal Hemp plant, the fibre having been exported from Sisal, in the Mexican part of Yucatán.

SITKA SPRUCE (*Picea sitchensis*, native of Pacific North America, introduced in the 19th century). 1895. Spruce from Sitka, town on Baranof Island, in Alaska.

SKIRRET (the root-vegetable *Sium sisarum*, native of Eastern Asia, introduced in the Middle Ages). 14th century, *skirwhitte*; 16th century, *skyrwort* (Turner 1548), *skirwurte* (Lyte 1578), *skirret*. English garbling, as if = *skire* + *white* (in reference to the white roots), of Old French *eschervys*, from medieval Latin *carvi*, from Arabic *karawiyā*, from Latin *careum*, from Greek *karon*,

197

karo, carraway. Both Carraway and Skirret are white-flowered umbelliferous plants.

SKULLCAP (*Scutellaria galericulata*). XVIII century. From the resemblance of the flower, when reversed, to a 'skull' or skull-cap, i.e. a form of close-fitting iron helmet worn in the 16th century.

SLOE (fruit of the Blackthorn, *Prunus spinosa*). Old English *slāh*, *slā* (genitive singular *slān*), for which there exist cognates in German, Dutch, Frisian, from an Indoeuropean base meaning 'bluish'.

SLOKE (the edible seaweed *Chondrus crispus*, Carragheen). XV century *slauk*. From the Irish *slabhac*, *sleabhac*, 'horn', describing the antler-shaped fronds.

SMALLAGE (*Apium graveolens*, Wild Celery; and the edible *Apium nodiflorum*, Fool's Watercress). XIII century. *small + ache*, 'celery', from the Old French *ache*, from Latin *apia*, plural of *apium*, from Greek *apion* (Dioscorides, etc.).

SNAKE'S-HEAD (*Fritillaria meleagris*, Fritillary). XIX century, Anne Pratt 1859.

SNAKEWEED (*Polygonum bistorta*). XVI century, Lyte 1578, translating the German *Natterwurz* which in turn translated the Late Latin *serpentaria* (for a different plant).

SNAPDRAGON (*Antirrhinum majus*, native of the Mediterranean region, introduced probably in the 16th century). XVI century (1573). The jaw-like flowers, suggestive of a dragon's head, can be opened and allowed to snap back into their natural posture.

SNEEZEWORT (*Achillea ptarmica*). XVI century, Gerard 1597, translating the German *Niesskraut* or the apothecaries' Latin *herba sternutatoria*, since *A. ptarmica* was identified with the Greek *ptarmike* (*ptarmikos*, 'causing to sneeze') of Dioscorides.

SNOWBALL TREE (*Viburnum opulus* var. *roseum*, Guelder Rose, introduced from Holland in the 16th century). XVIII century.

From the snowball-like clusters of sterile flowers. See *Guelder Rose*.

SNOWBERRY (*Symphoricarpos albus*, from North America, introduced 1730). XIX century. Descriptive of the pendent fruits.

SNOWDROP (*Galanthus nivalis*, probably introduced from Europe in the 16th century). XVII century (1664). Both from the white of the hanging flowers and their blossoming often through the snow. A translation of the German *Schneetropfen*, the earliest known vernacular name, recorded first by Charles de l'Écluse in 1583, *G. nivalis* not having been described by any botanist or other writer until 1571. It was taken to be an early blossoming kind of Snowflake (*leucoium bulbosum*), and was therefore named *leucoium bulbosum praecox* (Gerard 1597), whose English name was 'Timely flowring Bulbus Violet'. Bulbous Violet did duty for half a century. In his *Garden Book*, 1659, Sir Thomas Hanmer still calls *G. nivalis* the 'Early White kind of Bulbous Violet'. Miller in his great *Gardeners Dictionary*, 1731, talks of 'Bulbous-violet or Snow-drop' – 'very common in most *English* Gardens, where it is preserved for its early Flowering, which is generally in *January*, when it often appears, though the Ground at that time be covered with Snow, and is one of the first Usherers in of the Spring.'

SNOWFLAKE (*Leucojum aestivum*, Summer Snowflake, Loddon Lily). XVIII century. A name given by William Curtis in his *Flora Londinensis*, 1798, to distinguish *Leucojum* from *Galanthus* – see *Snowdrop*, above.

SOAPWORT (*Saponaria officinalis*, Bouncing Bett, probably introduced from Europe in the Middle Ages). XVI century, Turner 1548, translating the medieval Latin *Saponaria* (*herba*), or the German *Seifenkraut*: 'it groweth in certeine gardines of Germany, but I never sawe it in England, therefore I know no englishe name for it. Howebeit, if we had it here, it mighte be called in english sopewurt or skowrwurt.' (In *The Grete Herball*, 1526, *S. officinalis* had been given several English names, including *Saponary*, *Fuller's grasse*, *Herbe Phylyp*, and *Crowsope*. As abroad, it seems to have been used in the medieval cloth industry.)

SOFT GRASS (*Holcus mollis*). XVIII century (1785). From the Latin *mollis*, 'soft', of the Linnaean name *H. mollis*.

SOLOMON'S SEAL (*Polygonatum multiflorum*). XVI century (1543). A translation of the medieval Latin *sigillum salomonis*, the seal being apparently one of the flowers, hanging like a seal from a document, rather than any marking of the root. (The Seal of Solomon was the pentacle, the star of five points symbolizing the five wounds of Christ, which had the power of putting demons to flight.)

SOPS-IN-WINE (*Dianthus caryophyllus*, Clove-pink, Gillyflower, native of Southern Europe, introduced in the Middle Ages). XVI century (1573). I.e. plant with flowers which are the pink colour of sops, pieces of bread, soaked or dipped in red wine, a favourite dish. Later, 18th century, a variety of red-fleshed apple, or apple with red veins in the flesh.

SORB (fruit of *Sorbus domestica*, the Service Tree; and the tree). XVI century. From the Old French *sorbe*, from Latin *sorbum*, fruit of the *sorbus*. *Sorb-apple* (i.e. sorb). XVI century, Turner 1548, translating the German *sorbapfel*. See also *Service*.

SORGHUM (the cereal and forage grass *Sorghum vulgare*, of African origin). XVI century, Gerard 1597. A Latinizing of French *sorgho*, from the Italian *sorgo*, from the medieval Latin *surgum, suricum*, from a Romance **syricum* (*gramen*), 'Syrian grass'.

SORREL (*Rumex acetosa*). XIV century. From the Old French *surelle, sorele*, diminutive of *sur*, 'sour' or 'acid'–'little acid plant'.

SOUTHERNWOOD (*Artemesia abrotanum*, Lad's Love, native of Southern Europe, introduced early in the Middle Ages). Old English *sútherne-wudu*, 'woody (plant) from the south'. See *Lad's Love*.

SOWBRED (*Cyclamen neapolitanum*, native of Central and Southern Europe, introduced in the 16th century). XVI century (c. 1550). A translation of the Medieval Latin *panis porcinus*, 'pig's bread', i.e. food eaten by wild boars and pigs in the woods.

SOW-THISTLE (*Sonchus oleraceus*, Milk-thistle). XIII century, *sugethistel*, thistle eaten by sows.

SOYA, (sauce made from soya-beans; and so the soya-bean plant *Glycine max*, native of China and Japan, introduced from the East Indies 1790). XVII century. From Dutch *soja*, from Japanese *shōyu*, from Chinese *shi-yow*, 'salt bean' + 'oil'.

SPANISH BAYONET (*Yucca aloifolia*, native of Mexico and the southern United States, introduced in 1696). XIX century. An American name, from the bayonet-shaped leaves, for a plant common in the former Spanish territories.

SPEARMINT (*Mentha spicata*, native of the mountains of Central Europe, introduced in the Middle Ages). XVI century, Turner 1568, who ascribes the name to the sharpness of the leaf, not to the spiked inflorescence (cf. the 16th century botanists' Latin names *mentha angustifolia spicata*, 'narrow-leaved spiked mint', *mentha acuta*, 'sharp-pointed mint').

SPEAR-THISTLE (*Cirsium vulgare*). XVIII century. A translation of the 16th century Latin name *carduus lanceolatus*, 'thistle armed with little spears', or 'lances'.

SPEARWORT (*Ranunculus lingua*, Greater Spearwort; *R. flammula*, Lesser Spearwort). XIV century. Describing the shape of the narrow leaves. The Stockholm Medical MS., of the 15th century, says that *R. lingua* 'is an herbe that men clepe sperewourt or launcelef', adding that it has 'scharpe lewys as a spere'.

SPEEDWELL (species of the genus *Veronica*, first recorded for *V. officinalis*, Fluellen). XVI century, Lyte 1578. The sense (for a medicinal plant) is 'prosper well', 'go on well', 'get well'. *V. officinalis* was reputed to be a strengthening and wound-healing plant, also good against coughs. See *Fluellen*.

SPHAGNUM (bog-moss, made up of species of the genus *Sphagnum*; and the plants). In the second sense, XVIII century; in the first sense, XIX century. English use of the generic name, Latin *sphagnum,* a moss on trees described by Pliny, which the German botanist J. J. Dillenius (1687–1747), Professor of Botany at Oxford, first used for the bog-mosses, in his *Historia Muscorum,* 1741.

SPIDER ORCHID (*Ophrys sphegodes*). XVIII century, *Spider Orchis,* from the second element of the older botanists' Latin name *Ophrys aranifera,* '*spider-bearing Ophrys*', in reference to the flowers which resemble plump spiders.

SPIDERWORT (species of the genus *Tradescantia,* native of North and South America; especially *T. virginiana,* introduced from Virginia in 1629). XVII century. Parkinson 1629, *Tradescant his Spiderwort.* From the zigzag sprawl of the plant, suggestive of spiders' legs. Earlier, Gerard 1597, for the lily *Anthericum liliago,* translating the German (*Erd*)*spinnerkraut.* See also *Tradescantia.*

SPIGNEL (*Meum athamanticum,* Meu, Baldmoney). XV century, *spigurnel.* Of unknown origin.

SPIKENARD (eared rootstocks of the Indian plant *Nardostachys jatamansi,* of the Valerian family, used in making perfume). XIV century. From the medieval Latin *spica nardi,* 'spike or ear of nard', from Latin *spicae nardi,* 'spikes of nard' (Pliny), reproducing Greek *nardostachus* or *nardoustachus* (*stachus,* 'ear of corn', etc.), *nardos* deriving from Hebrew *nērdh,* cognates of which are Accadian *lardu* and Sanskrit *náladam.*

SPINACH (*Spinacia oleracea,* native of S.W. Asia, introduced in the 16th century – Turner 1568: 'an herbe lately found and not long in use'). XVI century, *spynnage, spinage,* from the Old French *espinage, épinach,* from Spanish *espinaca,* from Spanish Arabic *isbīnākh,* from Arabic *isbānakh*; from Persian *aspanākh,* with the same meaning.

SPINDLE-TREE (*Euonymus europaeus*). XVI century, Turner 1548, 'it maye be called in English Spyndle tree,' translating the German *spinnelboum, Spindelbaum,* a name going back to the 9th century. The wood was cut into spindles for the spinning of yarn.

SPLEENWORT (ferns of the genus *Asplenium*, especially *A. trichomanes*, Common Spleenwort). XVI century, Lyte 1578, translating the German *Miltzkraut*, itself an equivalent for medieval Latin *asplenium*, from Latin *asplenon*, Greek *asplenos*, *asplenon*, a fern used medicinally for the spleen (Greek *splen*).

SPRUCE (*Picea abies*, Norway Spruce, native of North and Central Europe, introduced early in the 17th century). XVII century, John Evelyn in his *Sylva*, 1670: 'for masts, etc. those [fir trees] of Prussia, which we call Spruce, and Norway . . . are the best.' From the country called Spruce, i.e. Pruce, or Prussia.

SPURGE (species of the genus *Euphorbia*; originally for *E. lathyrus*, Caper Spurge, native of Southern Europe, introduced in the Middle Ages). XIV century. From the Old French *espurge*, 'the purging herb' (medieval Latin *purgatoria*), the fruits, or seeds, having been taken as purgative pills. Cf. German *Scheisskraut*, Dutch *schijtkruid*. *E. lathyrus* was the medieval *cataputia* (Latinization of Italian *cacapuzza*, 'shit stink'), the 'catapus' which grew in the garden in Chaucer's *Nonnes Preestes Tale*, providing one of the 'laxatyves', pressed on Chantecleer by his wife, the fussy hen Pertelote.

SPURGE LAUREL (*Daphne laureola*). XVI century, Gerard 1597. From the resemblance of *D. laureola* to a large spurge.

SPURREY (*Spergula arvensis*, Corn Spurrey). XVI century. From the Dutch *spurie* (West Frisian *sparje*), from which the Flemish botanist Mathias de l'Obel seems to have coined the Latinized name *spergula* (1571). Cf. Lyte 1578, who gives the English name *francke*, adding 'It is also called in English Spurrie, and so it is in Frenche and Douch: whereof sprang the Latin name *Spergula*, unknowen of the Apothecaries, and the oldest writers also.' The cultivation of Spurrey as a forage crop was introduced from the Low Countries in the 16th century.

SQUASH (fruit of species of *Cucurbita*, pumpkins, from North America). XVIII century. A shortening of *askútasquash* in the Narraganset and Natick language of Massachusetts, meaning '(fruits) eaten green', or 'raw'.

SQUILL (the medicinal Sea Onion, *Urginea maritima*, native of Mediterranean Europe, introduced in the late Middle Ages; also *Scilla verna*, Spring Squill, and *S. autumnalis*, Autumnal Squill). In the first sense, xIV century; in the second sense, xVIII century. From Latin *scilla*, *squilla*, from Greek *skilla*, with the same meaning, the bulbs of *Urginea maritima* having been used by the Greeks medicinally, in rites of purification, and as a charm to protect houses from evil spirits.

SQUINANCY-WORT (*Asperula cynanchica*) xVIII century. Plant for healing squinancy, i.e. tonsillitis. *A. cynanchica* was made into an astringent gargle.

SQUIRREL-TAIL GRASS (*Hordeum marinum*). xVIII century. From the long-awned spike. A name recorded from the Isle of Thanet by William Curtis (1746–1799) in his great *Flora Londinensis*.

SQUITCH (the grass *Agropyron repens*, Couch, Quitch, Scutch, Twitch). xVIII century. A variant of *Couch*, q.v.

STAG'S HORN (*Rhus typhina*, native of North America, introduced in the early decades of the 17th century). xVIII century, *Stag's horn-tree*. From the thick pubescence of the young twigs, suggesting the velvet of growing horns.

STAG'S HORN MOSS (*Lycopodium clavatum*, Common Clubmoss). xVIII century. A name first recorded by the German-English botanist J. J. Dillenius (1687–1747), Professor of Botany at Oxford, in his pioneering *Historia Muscorum*, 1741. From the antler-like branches.

STAR ANISE (*Illicium verum*, native of China, introduced 1790; and its fruits used as a spice). xIX century. *Star*, from the star-like fruits, + *anise* (*Pimpinella anisum*), from their strong scent and flavour of anise (q.v.).

STAR OF BETHLEHEM (*Ornithogalum umbellatum*). xVI century. The inflorescence and white flowers suggest a formalized, pointed Star of Bethlehem, hanging over the birthplace of Christ.

STAR THISTLE (*Centaurea calcitrapa*, native of Central and Southern Europe, probably introduced by herbalists in the 16th century). XVI century, Lyte 1578, translating the Dutch *sterre distel* or the apothecaries' Latin *carduus stellatus*. From the star-like arrangement of the long sharp-pointed bracts.

STEPHANOTIS (*Stephanotis floribunda*, Madagascar Jasmine, native of Madagascar, introduced in 1839). XIX century. English use of the botanical Latin, from the Greek adjective *stephanotis*, 'fit for a wreath' or 'crown' (*stephanos*).

STINKHORN (the fungus *Phallus impudicus*). XVIII century. Ray (1724): 'This is known to all our Countrey People by the Name of Stinkhorns.' From the smell and the resemblance to a horn, i.e. an erect member.

STITCHWORT (*Stellaria holostea*, Greater Stitchwort). XIII century. A plant for the stitch in the side. Gerard 1597: 'They are woont to drinke it in wine with the powder of Acornes, against the paine in the side, stitches, and such-like.'

STOCK (*Matthiola incana*, native of the Mediterranean region, introduced in the late Middle Ages or the early 16th century). XVII century, short for *Stock-gilliflower* (1530), i.e. a gilliflower (q.v.) with a woody stock or shoot. See also *Brompton Stock, Ten Weeks Stock, Virginian Stock*.

STONECROP (various species of the genus *Sedum*; but originally *Sedum acre*, Wall-pepper). Old English *stāncrop*, 'that which is cropped, cut, gathered, off stone'.

STONE PARSLEY (*Sison amomum*). XVI century, Turner 1548, translating Greek (and Latin) *petroselinon*.

STONE PINE (*Pinus pinea*, native of the Western Mediterranean region, introduced in the 16th century). XVIII century, Miller in the seventh edition of his *Gardeners Dictionary* 1759. From the French *pin de pierre* (usually *pin pignon*).

STORAX (perfumed resin of the small tree or shrub *Styrax officinalis*; and the tree, native of the Mediterranean region, introduced in the late 16th century). In the first sense, xiv century; in the second sense, xvii century. English use of the Latin *storax*, from Greek *sturax*, 'storax gum', 'storax tree', from the Hebrew *tzori*, gum of the terebinth and mastic.

STORKSBILL (*Erodium cicutarium*). xvi century, Turner 1562, translating the German *Storchenschnabel*. From the long beak of the carpel. See also *Geranium*.

STOUT (dark brown beer or porter, qqv.). In the present sense, xix century; earlier, with the meaning 'strong beer', xvii century. Shortened from *stout beer*, i.e. a beer which is stout or strong rather than thin.

STRAWBERRY (*Fragaria vesca*, fruit and plant; and so the other species and varieties of Strawberry). Old English *strēawberige*, (*straw* + *berry*), in which *strēaw* may have its other Old English meaning of 'chaff' or 'mote', in reference to the chaff-like achenes which protrude over the surface of the strawberry.

STRAWBERRY TREE (*Arbutus unedo*). xvi century, Turner 1548; the tree having 'a fruite lyke a strawberry, wherefore it may be called in english strawberry tree'.

STURMER, STURMER PIPPIN (variety of dessert apple). xix century. Apple raised in Dillistone's nursery garden at Sturmer in Essex, near Steeple Bumpstead, introduced in 1843.

SUCCORY (*Cichorium intybus*). xvi century (1533). From the Dutch *suyckereye*, of Low German *suckereye*, forms of the Dutch and German derivatives of *cichoreum*. See *Chicory*.

SUGAR (cane-sugar from the juice of the grass *Saccharum officinarum*, which originated probably in the South Pacific region). xiii century, *suker*. From the Old French *çucre*, from Italian *zucchero*, from medieval Latin *succarum*, from Arabic *sukkar*, traceable through Persian and Pali to Sanskrit *śárkarā*, 'grit' (describing the look of granulated or powdered sugar).

SUGAR-BEET (form of *Beta vulgaris* ssp. *vulgaris*). xix century (1831). Major European source of sugar.

22. Wild Strawberry (Fuchs, *De Historia Stirpium*, 1542)

SUGAR-CANE (the grass *Saccharum officinarum*. See *Sugar*, above).
XVI century. A translation of the French *canne à sucre*.

SULPHUR TUFT (the fungus *Hypholoma fasciculare*). XIX century. From the sulphur-yellow cap, stipe, and flesh.

SULPHURWORT (*Peucedanum officinale*, Hog's Fennel). XVI century, Lyte 1578, translating the German *Schwefelwurtz*, on account of the sulphurous smell of the root, the sap of which was celebrated as a medicine against wind, cramps, toothache, etc. Also *Sulpherweed*, 19th century (Anne Pratt).

SUMACH (originally the dried leaves, used in tanning, e.g. of morocco leather, of the shrub or small tree *Rhus coriaria*, native of the Mediterranean region; also the plant; and other species of *Rhus*, especially *R. typhina*, Stag's-horn Sumach, from North America, introduced in 1629). XIV century, in the first sense; XVI century, in the second sense. From the Old French *sumac*, from the Arabic *summāq* (the shrub), from Syriac *summāq*, 'red' (*R. coriaria* furnishes a red dye).

SUNDEW (*Drosera rotundifolia*). XVI century, Lyte 1578, translating the apothecaries' Latin *ros solis*: 'This herbe is of a very strange nature and marvelous: for although that the Sonne do shine hoate, and a long time thereon, yet you shall finde it always full of little droppes of water: and the hoater the Sonne shineth upon this herbe, so muche the moystier it is, and the more bedewed, and for that cause it was called *Ros Solis* in Latine, which is to say in Englishe, The dewe of the Sonne, or Sonnedewe.'

SUNFLOWER (originally the Turnsole, *Heliotropum europaeum*, native of S.E. Europe, introduced in the 16th century; then *Helianthus annuus*, the Common Garden Sunflower, introduced from North America at the close of the 16th century). In the first sense, XVI century, Turner 1562, 'Because it turneth the leaves [i.e. its flowers, or flower-spikes] about wyth the sonne, it is called Heliotropion, that is, turned wyth the sonne, or sonne flower' – *heliotropion* having been the Greek name (Dioscorides).

In the second sense, for *Helianthus annuus*, with its heliotropic flowerheads, Gerard 1597, translating the botanists' Latin *flos solis*, or *solis flos peruvianus*, 'sun flower of Peru'.

SWEDE (*Brassica napus* var. *napobrassica*, Swedish Turnip, Rutabaga). Early XIX century. Introduced from Sweden in 1781.

SWEEPS, CHIMNEY-SWEEPS, SWEEP'S BRUSHES (*Luzula campestris*). XIX century. From the dark colour and shape of the flower clusters on stiff stems. Cf. the equivalent German names *Schlotfeger*, *Schornsteinfeger*.

SWEET BRIAR (*Rosa rubiginosa*). XVI century, Turner 1538, 'Swetebrere aut Eglentyne'. See also *Briar*.

SWEET CICELY (*Myrrhis odorata*). XVII century. From the medieval Latin *seseli*, from Greek *seseli*, an umbelliferous plant used in medicine, which was confounded with the girl's name Cicely. *M. odorata*, marked by the sweet scent of its leaves, was equated both with the *seseli* and the *murrhis* of Dioscorides. The spelling *Cicely* for *seseli* first appears in Gerard 1597.

SWEET FLAG (*Acorus calamus*, native of Southern Asia, introduced from Turkey in 1557). XVIII century; earlier *sweet smelling Flagge* (Parkinson 1640). A translation of the 16th century botanists' Latin *calamus aromaticus* (Pliny's *calamus odoratus*), from the Greek *kalamos aromatikos*, the rootstock having a pungent scent like that of orange peel (it was sold by apothecaries as *Rhizoma Calami*, one of the ancient medicines of the world, used also in T'ang China and in early – and modern – India).

SWEET GALE (*Myrica gale*, Bog Myrtle). See *Gale*.

SWEET PEA (*Lathyrus odoratus*, native of southern Italy, introduced in 1700). XVIII century (1732); earlier *Sweet-scented Pea* (1728).

SWEET POTATO (tubers of *Ipomoea batatas*, native of tropical America, introduced into Spain by Columbus in 1493, and available in London by 1577). These tubers were the first to be called Potato (1555 *botata*; 1577 *patata*), from the Spanish *patata*, from the Arawak name *batata*. Our potatoes, the tubers of *Solanum tuberosum*, introduced into Spain very much later, were distinguished by Gerard 1597 as 'Virginian potatoes'. These soon becoming the commoner kind, the tubers of Columbus's introduction needed eventually to be distinguished, in their turn, as the Sweet Potato (from their sugary taste), a name first recorded, according to the *Oxford English Dictionary*, in 1775. See also *Potato*.

SWEET SULTAN (*Centaurea moschata*, native of the Near East, introduced by 1629). XVIII century, first occurring, according to the *Oxford English Dictionary*, in James Gardiner's translation (1706) of the Latin poem on gardens by René Rapin (1621–1687):

Sweet-Sultans nam'd from the Byzantine King

– i.e., named from the Sultan of Turkey (cf. the French name, *Fleur du Grand Seigneur*). No doubt the rich flower suggested a turban worn by the Grand Turk. *Sweet*, from the honey scent of the flower.

SWEET WILLIAM (*Dianthus barbatus*, native from Southern Europe to Southern Russia, introduced by 1573). XVI century (1573), though earlier (1562) for the Wallflower (*Cheiranthus cheiri*). *Sweet*, i.e. scented.

SWINE'S CRESS (*Coronopus squamatus*, Wart-cress). XVI century. Grubbed up by pigs, a cress fit only for pigs.

SYCAMORE, SYCOMORE (*1. Ficus sycamorus*, the Near Eastern 'sycomore tree' of the Bible; *2. Acer pseudoplatanus*, native of the mountains of S.W. Asia and Central and Southern Europe, introduced into Scotland probably in the 15th century, into England c. 1565). In the first sense, XIV century; in the second sense, XVI century, Gerard 1597 (Gerard having known that the Sycamore was properly the biblical tree, and that this new maple had no right to the name, which was given to it because its leaves were similar to those of the Bible fig, and because both trees gave a dense shade). From the Old French *sicamor*, from the Latin *sycomorus*, from the Greek *sukomoros*, i.e. *sukon*, 'fig', + *moron*, 'mulberry', 'fig tree with mulberry-like leaves'. (The *Oxford English Dictionary* quotes, as its first record of Sycamore for *A. pseudoplatanus*, the speech by Boyet in Shakespeare's *Love's Labour's Lost*, 1598, which begins

> Vnder the coole shade of a Siccamore,
> I thought to close mine eyes some halfe an houre.

The priority, however, should go to Gerard's *Herbal* published the year before, in 1597, from which Shakespeare was evidently borrowing. Gerard calls the tree 'a stranger in England' – 'only it groweth in walkes and places of pleasure of noble men, where it especially is planted for the shadowe sake, and under the name of Sycomore tree.')

SYRINGA (*Philadelphus coronarius*, Mock Orange, native of S.E. Europe and S.W. Asia, introduced at the close of the 16th century). XVII century, Evelyn 1664. English use of the 16th century botanical Latin *syringa* (Latin *Syringa*, 'the nymph changed into a reed'), from Greek *surigx*, 'a reed' or 'pipe'. Both *P. coronarius* and the Lilac (*Syringa vulgaris*) were grouped together as *syringa*, or Pipe-tree, on account of their pithy stems. Parkinson 1629 calls lilac 'the blew Pipe tree', and *P. coronarius* 'the white Pipe tree'. So the continued use of syringa for the Mock Orange is a legitimate survival, not a mistake of the botanically uneducated.

The equally charming *Seringa* of the 18th century (1740) comes from French *seringat* (1718), from the same botanical Latin.

T

TAMARIND (pod of *Tamarindus indica*, native of South Asia and tropical Africa; and the tree). XVI century. From medieval Latin *tamarindus*, *tamar Indi*, from Arabic *tamr hindī*, 'date of India'.

TAMARISK (*Tamarix gallica*, native of Southern Europe and North Africa; *T. anglica*, native of western France, Spain and Portugal. Tamarisks are said to have been introduced by Edmund Grindal [1519?–1583], Archbishop of Canterbury). XV century. From the Late Latin *tamariscus*, from Latin *tamarix*, from Hebrew *tamar*, 'palm tree'.

TANG (seaweeds washed ashore). XVI century. A word of Norse origin found also in Norwegian, Danish, Faroese, Swedish, and German.

TANGELO (American hybrid of the Tangerine, *Citrus nobilis* var. *deliciosa*, and the Pomelo, or Grapefruit, *Citrus paradisi*; the fruit, and the tree). XX century. A hybrid word, *tang(erine)* + *(pom)elo*.

TANGERINE (*Citrus nobilis* var. *deliciosa*, native of Cochin-China). XIX century. Orange imported from Tangier, in Morocco.

TANGLE (seaweeds, and especially the stalked species *Laminaria digitata*). XVI century. From the Old Norse *thöngull*, 'Tang stalk'. See *Tang*.

TANSY (*Tanacetum vulgare*). XV century. From the Old French *tanesie*. *T. vulgare* was known to 16th century apothecaries both as *tanacetum* and *athanasia*. The names are apparently related, though it is not clear which of the two is the source of *tanesie*. Possibly *tanacetum* is a Latinization of *tanesie*, which, with loss of

212

the initial *a*, may derive from the Greek *athanasia*, 'immortality', which also had the meaning 'elixir' or 'antidote'. The reference would be either to *T. vulgare* as a medicinal herb or to the durability of its flowerheads and stems and the fact that *T. vulgare* is extremely persistent once placed in a garden. See also *Everlasting*.

TAPIOCA (flour from the root of the Cassava, *Manihot esculenta*, native in Brazil). XVIII century. From the Portuguese *tapioca*, from the Tupi-Guarani *typyóca*, meaning 'juice-fibre-gone', i.e. what is left after crushing the cassava root to get rid of the fibre and the poisonous juice.

TARE (the cultivation weed *Vicia hirsuta*, Hairy Tare). XIV century, as the seed of *V. hirsuta*; XV century, as the plant; described by Fitzherbert in *The boke of husbandrie*, 1523, as 'terre', the 'worste wede' of corn. A word of obscure origin.

TARO (*Colocasia esculenta*, originally from the East Indies). XVIII century. Hawaiian *taro*, first recorded by Captain Cook, 1779.

TARRAGON (*Artemesia dracunculus*, native of Asia; introduced from Southern Europe early in the 16th century). XVI century. From Spanish *taragoncia* (modern Spanish *taragona*), from the Arabic *tarkhūn*.

TEA (leaves of varieties of *Thea sinensis*, native of China). XVII century, as *tay*, *tee*, *tea*, *thea*, via Dutch *tee*, from the Chinese of Amoy, *t'e*, from Mandarin Chinese *ch'a*.

TEAK (the tree *Tectona grandis*, native of India and Malaya). Late XVII century. From the Portuguese *teca*, from the Malayalam *tēkka*.

TEA ROSE (*Rosa* × *odorata*, forms of which were introduced from Chinese gardens 1789–1845). 1850. Shortened from *Tea-scented Rose*, *Tea-scented China* (*Rose*).

TEASEL (*Dipsacus fullonum*). Old English *tǣsel*, from *tǣsan*, 'to card', 'tear', 'tease', + the suffix *-el*, 'that which teases', i.e.

the spiny head of the anciently introduced subspecies of *D. fullonum*, used for teasing cloth.

TEN WEEKS STOCK (*Matthiola incana* var. *annua*, native of Southern Europe, introduced by 1731). XVIII century, Miller 1731, 'The small annual Stock-gilliflower will produce Flowers in about ten Weeks after sowing (which has occasioned its being called *The ten Weeks Stock*).' See *Stock*.

TEREBINTH (the tree *Pistacia terebinthus*, native of the Mediterranean region). XIV century. From the Old French *térébinthe*, from Latin *terebinthus*, from Greek *terebinthos*. *P. terebinthus* was the original source of turpentine, q.v.

THALE-CRESS (*Arabidopsis thaliana*). XVIII century. Translation of the Linnaean name *Arabis thaliana*. *A. thaliana* was described in the 16th century by the German botanist Johannes Thal (1542–1583).

THISTLE (prickly plants of various genera of *Compositae*). Old English *thistel*, with similar names in other Germanic languages, a diminutive of a Germanic **thihsta*, from an Indoeuropean base meaning to prick.

THORN (*Crataegus monogyna* and *C. oxyacanthoides*, Whitethorn, Hawthorn). Old English and Old Norse *thorn*, i.e. 'the bush with thorns'.

THORN-APPLE (*Datura stramonium*). XVI century, Lyte 1578, *Thorn apple*, *Thornie apple*, translating the botanists' Latin *pomum spinosum*, referring to the large thorny capsules.

THOROW-WAX (*Bupleurum rotundifolium*, Hare's Ear). XVI century, Turner 1548, translating the German *Durchwachs*, 'because the stalke waxeth thorowe the leaves'.

THRIFT (*Armeria maritima*). XVI century. Apparently meaning that which thrives, is evergreen.

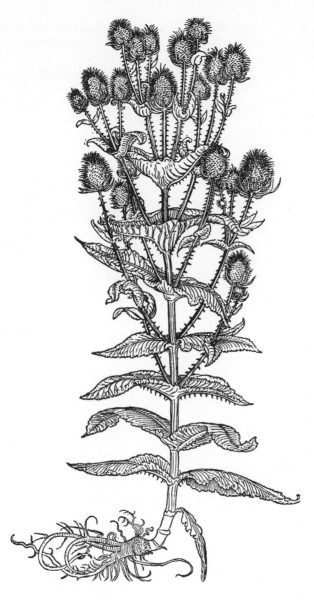

23. Teasel (Fuchs, *De Historia Stirpium*, 1542)

THROATWORT (*Campanula trachelium*). XVI century, Lyte 1578, translating the German *Halskraut*, the plant having been used as a source of astringent for sore throats.

THYME (species of the genus *Thymus*, especially the Garden Thyme, *T. vulgaris*, and the Wild Thyme, *T. serpyllum*). XV century. From Old French *thym*, from Latin *thymum*, from Greek *thumon*, 'that which is included in a sacrifice', from the verb *thuein*, 'to make a burnt offering'.

TIGER-LILY (*Lilium tigrinum*, native of China and Japan, introduced in 1804). XIX century. The first record in the *Oxford English Dictionary* is from *Our Village* (1824) by Miss Mitford.

TIMOTHY GRASS (*Phleum pratense*). XVIII century. After Timothy Hanson, American farmer who introduced seed of *P. pratense* from New York to Carolina, c. 1720.

TISTY-TOSTY (ball of cowslip flowers). XIX century. From the exclamatory 'hey tisty-tosty' (16th century), spoken presumably in the game in which girls tossed cowslip-balls to one another. Cf. Herrick's poem beginning

> I call, I call, who do ye call?
> The Maids to catch this Cowslip-ball.

TOADFLAX (*Linaria vulgaris*). XVI century. Turner 1548, translating the German *Krottenflachs*, i.e. a wild, useless flax, a flax for toads.

TOAD-RUSH (*Juncus bufonius*). XVIII century. From the *bufonius*, 'to do with toads', of the botanists' Latin name. Cf. the 16th century German *Krottengrass*, 'toad-grass' – as if a squat, useless rush, growing in toad's quarters.

TOADSTOOL (a stool-shaped fungus). XV century. (14th century, *tadstole*.) Cf. the toad names above.

TOBACCO (dried leaves of *Nicotiana tabacum*, first imported c. 1573; and the plant, native of tropical America, introduced late in the 16th century). XVI century. From Spanish *tabaco*, adapta-

24. Tobacco (Besler, *Hortus Eystettensis*, 1613)

tion of a Taino word for a roll of tobacco wrapped in a maize leaf; one of the words encountered in Haiti by Columbus.

Tom Putt, Tomput (West Country variety of apple). xx century. Said to be the name of a Somerset clergyman.

Tomato (*Lycopersicon esculentum*, native of western South America, introduced at the end of the 16th century). xvii century, *tomate* (from the mid 18th century, *tomato*). From the Spanish *tomate*, adaptation of the Nahuatl (Aztec) name *tomatl*, 'swelling fruit'.

Toothwort (*Lathraea squamaria*). xvi century, Gerard 1597, translating the German *Zahnwurtz*. The hard scale-leaves were likened to teeth, and *L. squamaria* was given against toothache.

Tormentil (*Potentilla erecta*). xv century. From the French *tormentille*, adapted from medieval Latin *tormentilla*, apparently a diminutive of *tormentum*, with the meaning of 'little torment plant'. The root of *P. erecta* was prescribed for griping stomach pains, and the pain or torment of toothache.

Touch-me-not (*Impatiens noli-tangere*). xvii century, in this sense, translating the *nòli-me-tangere* of 16th century botanists, which was taken from the words of the risen Christ to Mary Magdalene, in the Latin of the Vulgate (*John* xx). The capsules spring open at a touch. Touch-me-not was used by Gerard 1597 for the Squirting Cucumber (*Ecballium elaterium*), that still more startling plant from the Mediterranean region, which was grown in doctors' and apothecaries' gardens.

Tower Mustard (*Turrita glabra*). xviii century. A translation of the botanists' Latin name *Turritis*, from the Latin *turritus*, 'tower-shaped', descriptive of the tall straight stem.

Townhall Clock (*Adoxa moschatellina*, Moschatel). xx century. Describing the cube-shaped terminal flowerhead, with faces like those of such a clock.

Tradescantia (species of the American genus *Tradescantia*, Spiderwort; of which the first, *T. virginiana*, was introduced c.

1618). XVIII century. English use of the botanists' Latin name *Tradescantia*, given in 1718 in honour of the gardener and naturalist John Tradescant the Elder (d. 1637 or 1638), who was responsible for the introduction into England of *T. virginiana*. See *Spiderwort*.

TRADESCANT'S HEART (variety of cherry, the Noble). XIX century; 17th century, 'Tradescant's Cherry', grown by John Tradescant, d. 1637 or 1638, in the Earl of Salisbury's gardens at Hatfield House. *Heart*, short for *heart-cherry* (Gerard 1597, Englishing *cerasus cordata*, 'heart-shaped cherry').

TRAVELLER'S JOY (*Clematis vitalba*, Old Man's Beard). XVI century, Gerard 1597, a plant 'decking and adorning waies and hedges, where people travell, and thereupon I have named it the Traveller's Ioie'.

TREACLE MUSTARD (*Erysimum cheiranthoides*). In this sense, XIX century. Earlier, the weed *Thlaspi arvense*, for which it was coined by William Turner 1548: *triacle Mustard*, i.e. mustard for a treacle or remedy (Greek *theriake*, 'antidote against a poisonous bite').

TREE (tall plant with a woody trunk and branches). Old English *trēo*, Old Norse *tré*, a word of the Germanic languages from an Indoeuropean **deru-*, **doru-*.

TREE OF HEAVEN (*Ailanthus altissima*, native of China, introduced by seeds from Nanking 1751). XIX century. A translation of *ailanthe*, 'tree-heaven', the name in Amboina, Indonesia, for a related species, *A. moluccana*.

TREFOIL (species of the genus *Trifolium*, Clover; and of related genera). XIV century. From the Anglo-Norman *trifoil*, from Latin *trifolium*, from Greek *triphullon*, clover, literally 'three-leaf'.

TRUFFLE (the underground fungus *Tuber melanosporum*, the Perigord Truffle; *T. magnatum*, the Piedmont Truffle; and the English *T. aestivum*). XVI century. From the French *truffe*, from

Perigordian *trufa*, which, by the Low Latin *tufera*, goes back to a dialect form of the Latin *tuber*, a truffle, literally, a swelling. For a *tuber*, as truffle, see Pliny, who describes how an official in Spain bit on a denarius inside a truffle and damaged his teeth; and Apicius, who gives recipes for cooking truffles in his *De re coquinaria*.

TULIP (species of the genus *Tulipa*, of which *T. gesneriana*, native of Armenia and Persia, was introduced into England c. 1577). XVII century, *tulip*, *tulipe* (Lyte 1578, *Tulpia*, *Tulipa*, *Tulpian*). A shortening of *tulipa*, or from Dutch *tulp*, from *tulipan*, the name recorded c. 1554 by Ogier Gheselin de Busbecq (1522–1592), Viennese ambassador to Suleiman the Magnificent, at Constantinople, from where Busbecq sent back the first seeds to reach the West. His *tulipan* was the Turkish *tulband*, 'a turban' (from Persian *dulban*), but Busbecq seems to have mistaken a description of the flower for its name (the Persian and Turkish for tulip is *lalé*), as if his informant had remarked only that tulips were like turbans.

TULIP-TREE (the North American *Liriodendron tulipifera*, introduced in 1663). XVIII century. A tree with tulip-like flowers.

TURK'S CAP (*Lilium martagon*, Martagon Lily, native of the European mountains, probably introduced, at the end of the 16th century). XVII century, in this sense. The name (Gerard 1597) was first used for the Tulip (q.v.).

TURMERIC (root, or powdered root, of *Curcuma longa*, native of India). XVI century. Apparently a garbling of French *terre-mérite*, from medieval Latin *terra merita*, 'proper earth', 'earth of merit'. The English name (Turner 1538) was also given to the root of Tormentil (*Potentilla erecta*).

TURNIP (tap-root of *Brassica rapa*; and the plant). XVI century. The second element is from the Old English *nǣp*, from Latin *napus*, turnip. Possibly the 'turn-neep' or 'turn-nep', 'the rounding nep', from the swollen round shape. An English word

25. Traveller's Joy, Old Man's Beard (Fuchs, *De Historia Stirpium*, 1542)

which went abroad into other languages with the turnip culture of the 18th century agricultural improvers.

TURNSOLE (*Heliotropum europaeum*, Heliotrope, native of S.E. Europe, introduced in the 16th century; earlier, the dyestuff from *Chrozophora tinctoria*, native of Mediterranean Europe, and the plant). In the first sense, XVI century, Lyte 1578; in the second sense, XIV century; in the third sense, XVI century, Lyte 1578. From Old French *tournesol*, from Italian *tornasole*, from Latin *tornare*, 'to turn', + *sol*, 'sun', 'plant whose flowers turn with the sun'.

TURPENTINE (earlier the resinous sap of the Terebinth, *Pistacia terebinthus*, native of the Mediterranean region). XIV century. From Old French *terbentine*, from Latin *terebenthina* (*resina*), from Greek *terebinthine rhetine*, 'resin of the terebinth'.

TUSSOCK (clump of sedge, rushes, or coarse grass). XVII century, in this sense; XVI century, a tuft of hair. Of unknown origin, related apparently to obsolete *tusk* (16th century), which had the same two meanings.

TUTSAN (*Hypericum androsaemum*). XV century. From the Anglo-Norman *tutsaine*, French *toute-saine* (still used for *H. androsaemum* in Normandy), 'all wholesome', 'all pure', the plant having been taken as the Agnus Castus (q.v.), the plant of chastity associated with the Agnus Dei, the Lamb of God. 'Agnus castus is an herbe that men clepyn totsane . . . the vertue of this herbe is this that he wylle gladly kepe men and women chast' (Stockholm MS. X90. Early 15th century).

TWAYBLADE (*Listera ovata*). XVI century, Lyte 1578, translating the German *Zweyblatt*, 'two leaf'.

TWITCH (*Agropyron repens*, Couch-grass). XVI century. A form of Couch, Quitch, qqv.

U

UGLI (the Tangelo, hybrid between *Citrus paradisi*, the Grape-fruit, and *C. nobilis* var. *deliciosa*, the Tangerine; and its fruit). XX century. A differentiated spelling of 'ugly', in reference to the soft, rumpled, ugly look of the fruit. See also *Grapefruit, Pomelo, Tangelo, Tangerine*.

UPAS, UPAS TREE (fabled Javanese tree of death). XVIII century. From the Malayan *ūpas*, poison. This tree of death was concocted and made known in an article in the *London Magazine* in 1783, by the Shakespearean critic and joker George Steevens (1736–1800). He based his Upas on earlier accounts of a real Malaysian tree, *Antiaris toxicaria*, which itself is now called the Upas.

USQUEBAUGH (whisky – generally Irish whiskey; distilled from barley). XVI century. From Irish, as in Gaelic, *uisge beatha*, 'water of life'. See also *Whisky*.

V

VALERIAN (*Valeriana officinalis*). XIV century. From the Old French *valériane*, from the medieval Latin *valeriana* (*planta*), i.e. plant from Valeria, Roman province between the Danube, the Rába, and the Dráva (western Hungary).

VANILLA (pod of the tropical American orchid *Vanilla fragrans*; flavouring from the pods; and the plant). XVII century (in the first sense). From the Spanish *vaynilla* (modern Spanish *vainilla*), diminutive of *vaina*, from Latin *vagina*, 'a sheathe' or 'scabbard'.

VEGETABLE (substantival use of the adjective). As a plant, in contrast to the animal and mineral, late XVI century; in our common sense of a garden plant for cooking and eating, not until the late XVII century, in which sense the *Oxford English Dictionary*'s earliest quotation is taken from *A Farmer's Letters to the People of England*, by the great agricultural writer Arthur Young, published in 1768: 'The cultivation of the newly discovered vegetables', etc. The adjective *vegetable*, 'that which has the nature of a plant' (14th century), comes from Latin *vegetabilis*, from *vegetare*, in Low Latin, 'to grow'; in classical Latin, 'to invigorate'; which comes in turn from *vegetus*, 'lively', 'active', from *vegere*, 'to quicken', 'excite'.

VEGETABLE MARROW (form of *Cucurbita pepo*, the Pumpkin, the original home of which is unknown). XIX century. The sense is marrow – the edible goodness inside – of a vegetable kind. The *Oxford English Dictionary*'s first record of Vegetable Marrow is dated 1816.

VEGETABLE OYSTER (*Tragopogon porrifolius*, Salsify, native of the Mediterranean region, introduced from France and Italy into English gardens c. 1700). XIX century. From the taste.

224

VENUS LOOKING-GLASS (*Specularia hybrida*). XVI century, Gerard 1597, giving his version of the German *Frauwenspiegel*, 'Lady's looking-glass', descriptive of the flower topping its long ovary.

VENUS'S NAVELWORT (garden species of the genus *Omphalodes*, including *O. verna*, introduced from Southern Europe in 1633). XVIII century. In the 16th century the name was given to the Pennywort, *Umbilicus rupestris*. In *Omphalodes* the seeds and flowers, not the leaves, suggest the navel. See *Navelwort*.

VERBENA (garden plants of the genus *Verbena*, especially forms of the South American *V. teucrioides*, introduced in 1837). XIX century, in this sense. English use of the Linnaean name, from Latin *verbena*, in plural *verbenae*, branches of olive, laurel, tamarisk, myrtle, cypress, used in religious ceremonies. See *Vervain*.

VERMOUTH (alcoholic drink flavoured with Wormwood). See *Wormwood*.

VERONICA (species of the genus *Veronica*, the Speedwells; also shrubs of the New Zealand genus *Hebe*, formerly included under *Veronica*). XVI century, ? from St Veronica. *Veronica* for the Speedwells is first recorded by the German botanist Leonhard Fuchs in his *De historia stirpium*, 1542, as the name then used by most herborists.

VERVAIN (*Verbena officinalis*). XIV century. From the Old French *verveine*, from Latin *verbena*. See *Verbena*.

VETCH (species of the genus *Vicia*, especially the farmer's weed *Vicia hirsuta*). XIV century. From the Anglo-Norman and Old Norman French *veche*, from the Old French *vecce*, from Latin *vicia*.

VICTORIA (variety of plum). XIX century. This famous plum, which enshrines the taste of bad Victorian cooking (stewed plums), was evidently named in honour of the Queen. The variety was found in a garden at Alderton, in Sussex. It was sold to a Brixton nurseryman named Denyer, who marketed it c.

1840, three years after Victoria's accession. It was long known as Denyer's Victoria.

VINE (*Vitis vinifera*, the Grape Vine, probably a native of the Caucasus, first grown in Britain in the Roman era). XIII century. From the Old French *vigne*, from Latin *vinea*, 'vine', 'vineyard', feminine of the adjective *vineus*, from *vinum*, 'wine'.

VIOLET (species of the genus *Viola*, subgenus *Viola*, particularly the Sweet Violet, *V. odorata*). XIV century. From the Old French *violette*, diminutive of *viole*, from the Latin *viola*, which was the name given to the Stock (*Matthiola incana*) as well as the Violet.

VIPER'S BUGLOSS (*Echium vulgare*). XVI century, Lyte 1578. *E. vulgare* was equated with the *ekhion* which Dioscorides recommended as a preventive and cure of snakebite, the seeds, as Dioscorides observed of the seeds of *ekhion*, resembling snake's heads. Herborists of the 16th century took *E. vulgare* to be a bugloss, on account of its hairy roughness, and knew it as *sylvestre Buglossum*. Lyte's name more exactly defined it. See also *Bugloss*.

VIRGINIA CREEPER (*Parthenocissus quinquefolia*, introduced from North America in 1629). XVII century. It was introduced from the Virginian colony.

VIRGINIAN STOCK (*Malcolmia maritima*, native of the Mediterranean region, introduced in 1713). XVIII century (1760). Called Virginian in error, *M. maritima* was known to the earlier 18th century gardeners as Dwarf Annual Stock, from its resemblance (in scent, flower, etc.) to the Stocks or Stock-gillyflowers (*Matthiola*). The Mediterranean origin of *M. maritima* is indicated in its French name, *giroflée de Mahon*, i.e. of Port Mahon, in the Balearic Islands. See also *Stock*.

26. Vine (Fuchs, *De Historia Stirpium*, 1542)

W

WALLFLOWER (*Cheiranthus cheiri*, native of the Eastern Mediterranean, introduced c. 1573). XVI century, Lyte 1578. From its common habitat.

WALL RUE (the fern *Asplenium ruta-muraria*). XVI century, Turner 1548. Englishing of medieval Latin *ruta muraria*, so called for its Rue-like fronds.

WALNUT (fruits of *Juglans regia*; and the tree, native of S.E. Europe eastwards, grown in Roman Britain). Old English *walh-hnutu*, 'foreign nut', i.e. nut from Gaul, the land of foreigners, of the Romans. A word borrowed late in English from the Low German.

WART-CRESS (*Coronopus squamatus*, Swine's Cress). XIX century. Englishing of the German *Wartzen Kress* (1586), or the 16th century botanists' Latin *nasturtium verrucarium* (*verruca*, 'a wart'). The round, pitted, clustered siliculae resemble warts.

WATERCRESS (*Nasturtium officinale*). XV century. The Old English names were *ēa-cerse*, *wielle-cerse*, 'stream cress'.

WATER SOLDIER (*Stratiotes aloides*). XVII century. Adapted from the Greek name of a plant in Dioscorides with which *S. aloides* was identified, the '*stratiotes* which grows in the waters' (*stratiotes*, 'a soldier'). 16th century apothecaries and botanists also used the Latin name *militaris* ('soldier').

WATER-VIOLET (*Hottonia palustris*). XVI century, Gerard 1597, translating either the botanists' Latin *viola aquatilis* or the German *Wasserveiel*, since *H. palustris* has flowers like those of the *Veiel*, i.e. the Stock, *Matthiola incana*.

228

WATTLE (various Australian shrubs and trees of the genus *Acacia*). XIX century. Shortened from the earlier 19th century *Wattle-tree*, tree whose branches were used like osiers for making wattles, hurdles, etc.

WAYFARING-TREE (*Viburnum lantana*). XVI century, Gerard 1597 – Gerard having supposed that the French name *viorne* (from Gallo-Roman *viburna*, plural, as singular, of Latin *viburnum*) meant something ornamenting the road (Latin *via*) or wayside; hence 'the shrub seen by wayfarers'.

WEED (useless and abundant plant). Old English *wēod*, an Old Saxon word (*wiod*), a form of which is still used in Low German.

WEIGELIA (species of the genus *Weigelia*, especially *W. floribunda* introduced from China in 1845). XIX century. English use of the generic Latin name given in honour of the German botanist Christian Weigel (1748–1831). (Also Weigela.)

WELD (*Reseda luteola*, Dyer's Rocket). XIV century. A Germanic name, cf. German *Wau*, Low German *Waude*.

WELLINGTONIA (*Sequoiadendron giganteum*, Big Tree, Giant Redwood, native of California, discovered and introduced in 1853). XIX century. English use of the generic Latin name *Wellingtonia* (*gigantea*), given to *S. giganteum* by the English botanist John Lindley soon after it was discovered, in honour of the Duke of Wellington, who had died the year before (1852). Tree and Duke were compared in loftiness. Americans preferred the name Washington Tree. See *Sequoia*.

WELSH POPPY (*Meconopsis cambrica*). XVIII century, 'Welsh, or Yellow Wild Bastard Poppy'. Miller's *Gardener's and Florist's Dictionary*, 1724. In the 17th century *M. cambrica* had been named *Papaver Cambro-Britannicum*. This native of Wales, S.W. England, western France and N.W. Spain was first known from Wales.

WHEAT (*Triticum aestivum*, anciently from Western Asia; and its grains). Old English *hwǣte*, with cognates in other Germanic languages. The original meaning is 'white'.

WHIN (*Ulex europaeus*, Gorse, Furze). XV century. From Old Norse **hvin* which is found in place-names, e.g. Whinburgh, in Norfolk. It is the name for *U. europaeus* in the Scandinavian areas of England, in the east and north.

WHISKY (alcoholic drink distilled from barley). XVIII century. Shortened from Gaelic *uisge beatha*, 'water of life'. See also *Usquebaugh*.

WHITEBEAM (*Sorbus aria*). XVIII century. But apparently from an Old English **hwīt bēam*. The leaves are white underneath, and the tree shows white when there is a wind (cf. 'wind-beat whitebeam' in Gerard Manley Hopkins' sonnet 'The Starlight Night'). Other Old English names for *S. aria* were *hwītingtrēow*, 'tree which whitens', *hāran wīthig*, 'hoary withy'.

WHITE PEPPER. See *Pepper*.

WHITETHORN (*Crataegus monogyna* and *C. oxyacanthoides*, Hawthorn, May-tree). XIII century. ? from an Old English **hwīt thorn*, in contrast to Blackthorn (q.v.), though the late appearance of Whitethorn in English and *Weissdorn* in German suggests that the source may be the medieval Latin *alba spina*, from which also comes the French *aubépine*.

WHITLOW-GRASS (*Erophila verna*). XVI century, Gerard 1597. *E. verna* was known to 16th century herbalists as *paronychia vulgaris*, from its use in curing a whitlow (Latin *paronychia*). Hence Gerard's coinage for a plant which was too inconspicuous to have acquired common names.

WHORTLEBERRY (*Vaccinium myrtillus*; and fruit. Bilberry, Hurtleberry). XVI century, Lyte 1578, recording the name used for the bilberries which grew on the hill-tops round his home, Lytes Carey House, Charlton Mackrell, in Somerset. The variants *whortleberry* and *whorts*, *hurtleberry* and *hurts*, are used in Somerset, Hampshire, Devon and Cornwall. The derivation is unknown. See also *Huckleberry*.

WILD SERVICE TREE (*Sorbus torminalis*). XVII century. See *Service, Service Tree*.

WILLIAM, WILLIAMS (variety of dessert pear). XIX century. Shortened from *Williams's Bon Chrétien*. The pear was raised c. 1770 by a schoolmaster at Aldermaston, Berkshire. His name was Stair, but the pear is called after a nurseryman at Turnham Green, Chiswick, who introduced it commercially. See *Bartlett, Bon Chrétien*.

WILLOW (species of the genus *Salix*). Old English *wilig*, a name with cognates in the other West Germanic languages.

WILLOW-HERB (species of the genus *Epilobium*). XVI century, Lyte 1578. An Englishing of the Latin name *salicaria (herba)*, used by 16th century herbalists – *salicaria*, 'because the leaves are like those of the willow, or because it grows among willows' (Fuchs, *De Historia Stirpium* 1542).

WINDFLOWER (*Anemone nemorosa*, Wood Anemone). XVI century. Coined by Turner 1551 for Greek *anemone (anemos*, 'wind', + the feminine patronymic suffix *-one*, 'daughter of the wind'), in the herborists' Latin *herba venti*, 'herb of the wind': he quoted Pliny's remark (about *Anemone coronaria*) that the anemone derived its name from the fact that 'the floure never openeth it selfe, but when the wynde bloweth'.

WINE (fermented drink from the grapes of *Vitus vinifera*, native probably of the Caucasus). Old English *win*, a word common, in various forms, to the Germanic languages, from a Germanic **winam*, and so from Latin *vinum*. For the ancient source of *vinum* and the related Greek *oinos* (and of related Semitic words for wine) one must look outside the Indoeuropean languages, presumably towards the source of the vine and the grapes – to a Caucasian language, in the region of Armenia and the ancient Khaldian kingdom of Urartu.

WINTER CHERRY (*Physalis alkekengi*, Chinese Lanterns, Alkekengi, native of S.E. Europe, introduced by 1548). XVI

231

century, Turner 1548, 'in Poticarie Latin Alkakengi, in english Alcakeng or wynter cheries'.

WINTERGREEN (*Pyrola media*). XVI century, Turner 1548, giving an English equivalent, as he says, for the German *Wintergrün*.

WISTARIA (species of the genus *Wisteria*, especially *W. sinensis*, native of China, introduced in 1818, and *W. floribunda*, from Japan, introduced in 1820). XIX century. From the English use of the Latin generic name *Wisteria* (the original spelling), given in honour of the Pennsylvanian physician Caspar Wistar (1761–1818).

WITCHES' BROOM (twiggy growth on birches and other trees caused by mites). XIX century.

WITCHES' BUTTER (kinds of jelly-like fungus, especially the greenish-black *Exidia glandulosa* on decaying wood, and the orange-yellow *Tremella mesenterica* on dead branches). XIX century.

WITHWIND (*Convolvulus arvensis*, Bindweed). Old English *withowinde*, *withewinde*, from Germanic **withiwindō*, meaning 'the string(-like), rope (-like) winding plant' (cf. Old English *-winde* for a winding twisting plant). Withwind is still the name spoken with feeling by many thousands of gardeners in southern England.

WITHY (willow shoot for making baskets, etc.; and various species of *Salix*, especially *S. viminalis*, Osier). Old English *wīthig*, in its various forms a common Germanic word, the Indo-european base of which means 'something pliant', 'something for tying' (as one ties a faggot) or 'plaiting'.

WOAD (*Isatis tinctoria*, native of Central and Southern Europe, anciently introduced and cultivated; and the dye from its leaves). Old English *wad*, a West Germanic word.

WOLF'S-BANE (*Aconitum vulparia*, native of Europe, introduced in the 16th century). XVI century, coined by Turner 1548 for the

27. Woad (Fuchs, *De Historia Stirpium*, 1542)

Latin *lycoctonum*, Greek *lukokton*, 'wolf-killing (plant)', mentioned by Galen.

WOOD-ANEMONE (*Anemone nemorosa*, Windflower). XVII century. An Englishing of the herborists' Latin *anemone nemorum*, 'anemone of the woods'. See *Anemone, Windflower*.

WOODBINE (*Lonicera periclymenum*, Honeysuckle). Old English *wudu-binde*, 'wood binder'.

WOODRUFF (*Asperula odorata*). Old English *wudu-rōfe*, i.e. *wood* + a word or plant name of unknown significance. The sense should be something that creeps or spreads.

WOOD SAGE (*Teucrium scorodonia*). XVI century. Descriptive of the sage-like leaves.

WOOD-SORREL (*Oxalis acetosella*). XVI century. From the late Middle Ages to the 17th century *O. acetosella* was commonly known in English as *sorrel de boys*. The other common name was *Alleluia* (from early blossoming at Easter, when Alleluia was sung on Easter Sunday and through Easter week after the Psalms), which is still the French name. *Sorrel de boys* was gradually superseded by its English translation, which also drove out *Alleluia*.

WORCESTER PEARMAIN (variety of apple). XIX century. Said to have been raised by a firm at Worcester, about 1873. See *Pearmain*.

WORMWOOD (*Artemisia absinthium*). XV century, in its present form (*wormewode*); it is not *worm* + *wood*, but from Old English *wermōd*, which is paralleled by West Germanic cognates (cf. modern German *wermut*, from which French *vermouth* is a late 18th century derivative). The significance is obscure. The use of the bitter wormwood against worms in the body seems to have been a consequence, rather than a cause, of the altered spelling.

WORT (a plant, especially a vegetable, or a plant used medicinally). Old English *wyrt*, a Germanic word with the basic meaning of 'root'. Cf. Old English terms for a vegetable garden, *wyrt-tūn*,

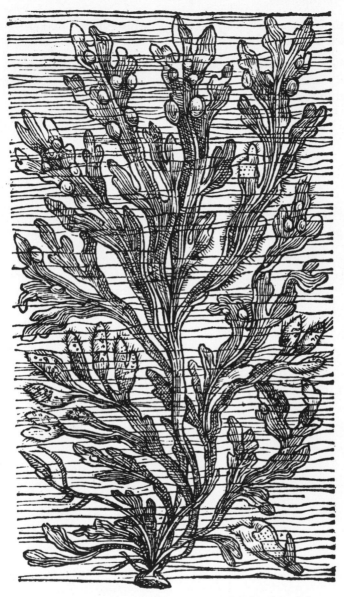

28. Wrack, Bladder Wrack (Lobel, *Kruydtboek*, 1581)

wyrt-geard (as *win-geard*, 'vineyard'). *-wort* is frequent in compound plant names, though these are often late fabrications by the 18th and 19th century botanists.

WOUNDWORT (name given to several vulnerary plants, now to species of the genus *Stachys*). XVI century, coined by Turner 1548 from the German (*heidnische*) *Wundkraut*, ' (heathen) wound-plant', *Senecio fluviatilis*, also known as Saracen's Consound (q.v.).

WRACK (seaweeds driven on shore, now, in particular, species of the genus *Fucus*). In this seaweed sense XVI century, a specialized application of *wrack* (= 'wrecked ship', 'wreckage'). From the Dutch *wrak*. ('Wreck', in the modern sense, which was also used from the 15th century for seaweed, derives from the similarly related Anglo-Norman *wrec*, a word from the Old Norse.) The basic Indoeuropean meaning is 'that which is driven'.

Bladder-wrack (*Fucus vesiculosus*), XIX century. From the little bladders or air vesicles on the fronds.

WYCH-ELM (*Ulmus glabra*). XVII century. *wych* = Old English *wice*, a wych-elm or other yielding tree, from a Germanic base meaning 'pliant', 'bending' (cf. the modern German *weich*, 'soft', 'pliant').

WYCH-HAZEL (*Hamamelis virginiana*, native of North America, introduced in 1812). XVIII century; but Wych-hazel was already applied in the 16th century to the Hornbeam (*Carpinus betulus*) and the Wych-elm (*Ulmus glabra*). See *Wych-elm, Hazel*.

Y

YAM (species of the tropical genus *Dioscorea*; and their tubers, first brought from Africa in the 16th century). XVII century. From the Portuguese *inhame*, possibly an African word, though its origin is unknown. If African, it is the first African word to have found its way into English. Forms earlier than *yam* were scarcely anglicized, e.g. *iniamo*, 1598.

YARROW (*Achillea millefolium*). Old English *gearwe*. The German name is *garbe*, the Dutch *gerwe*. The origin and significance of Yarrow are unknown, though it has been suggested that *gearwe* is related to the verb *gierwan*, 'to prepare', the sense being 'that which is prepared' or 'ready', i.e. 'the healing plant', *A. millefolium* having been highly reputed for healing wounds (and as a herb against enchantment).

YEAST (substance of fungal cells used to promote fermentation). Old English *gist*, **giest* which, with related names in other Indoeuropean languages, derives from an Indoeuropean base **yes-*, meaning 'to foam' or 'boil'.

YELLOW ARCHANGEL. See *Archangel*.

YELLOW INGESTRIE (apple variety). XIX century. Apple raised by the great horticulturalist Thomas Andrew Knight (1759–1838), and named after Ingestrie Hall, Staffordshire, seat of the agricultural improver Earl Talbot (1777–1849).

YELLOW LOOSESTRIFE (*Lysimachia vulgaris*). XVI century, Turner 1548, translating the herbalists' Latin *Lysimachiam luteam*: 'it groweth by the Temes syde beside Shene, it may be called in englishe yealow Lousstryfe or herbe Wylowe.' See also *Loosestrife*.

237

YELLOW TANG (the seaweed *Ascophyllum nodosum*, Knotted Wrack). XIX century. (Scotland). See *Tang*.

YELLOW-WORT (*Blackstonia perfoliata*). XVIII century. A wort with yellow flowers. *B. perfoliata* was used medicinally as a bitter herb.

YEW (*Taxus baccata*). Old English *īw*, *ēow*. There are cognate names in other Germanic languages and in the Celtic languages.

YGGDRASIL (in Norse mythology the ancient evergreen tree of fate, under which the gods sit in council and which upholds the universe; a yew, or an ash). XVIII century. Old Norse *yggr*, 'the Terrifier' (one of the names of Odin), + *drasill*, 'a horse', i.e. the tree which carried the hanged Odin as Lord of the Gallows.

YLANG-YLANG (the Malayan and Philippine tree *Cananga odorata*; and the perfume derived from its flowers). XIX century. From *alang-ilang*, the name of the tree and perfume in Tagalog, the national language of the Philippines.

YORK AND LANCASTER ROSE (*Rosa damascena* var. *variegata*). XVII century. With its white and red in one flower, the rose taken to symbolize the reconciliation of the royal Houses of York and Lancaster after the Wars of the Roses.

YORKSHIRE FOG. See *Fog*.

YUCCA (species of the North and Central American genus *Yucca*). XVII century. Originally a Carib word for the Cassava (*Manihot esculenta*), in which sense it reached English, from the Spanish *yuca*, in the 16th century.

Z

ZINNIA (species of the genus *Zinnia*, especially *Z. elegans*, introduced from Mexico in 1796). XVIII century. English use of the botanical Latin name given by Linnaeus in honour of the German botanist Johann Gottfried Zinn (1727–1759).